An Illustrated History of the U.S. Navy's Use of Law Enforcement-Style Badges
(Master-At-Arms, Police Petty Officers, Shore Patrol and others)

* * * * *

A military history reference book

DANIEL SMITH BOOKS
DANIELSMITHBOOKS.NET
MILITARY HISTORY BOOKS AND NOVELS

Cover design by Daniel D. Smith, Jr.

Printed in the United States of America

Smith, Daniel D., 1944-
An Illustrated History of the U.S. Navy's Use of Law Enforcement-Style Badges (Master-At-Arms, Police Petty Officers, Shore Patrol, and others)

Non-fiction by the author:

- U.S. Navy Tailor-Made Dress Blues, Liberty Cuffs, and Sailor Folk Art
- U.S. Navy Sailor's Memorabilia (Web Belt Buckles, Unit Plaques, Lighters and Ashtrays)
- The Medal of Honor and the Battles for Chattanooga

Fiction:

- The Biltmore's Mona Lisa

ISBN: 978-0-692-64374-7

DanielSmithBooks, LLC
DSBooks@epbfi.com
Chattanooga, Tennessee
United States of America

I am especially proud of my 26-year military career. Having completed a second career in the civilian work world, the thing I missed the most from my military days was the special camaraderie I felt during my years in the military, both active duty and reserve duty.

There is nothing in the civilian work force that can approximate the bonding that occurs in the wardroom, ready room, or foxhole. Military personnel in those environments put up with much hardship – long hours, stressful working conditions, danger to personal safety, separation from loved ones, and more. However, because they are all in it together, they get through it. This mutual self-sacrifice, teamwork, and covering each other's six contribute to individual bonding, unit cohesion, and, ultimately, the camaraderie...

Other than perhaps the professions of law enforcement, fire fighters, and emergency medicine (notice the common denominator), it would be difficult to find a civilian occupation that approximates the conditions of the [intense military training or battle situations]. It follows therefore that finding the military version of camaraderie in a civilian occupation is almost impossible.

...Because of the high frequency of duty station rotation in the service, it is important that the system accommodates these transitions. Getting onboard and up-to-speed quickly contributes to the automatic "welcome to the club" experience that awaits most military personnel when they report to a new duty station. This contributes to the camaraderie of military service...

—Thomas Wolfe
Out of Uniform: Your Guide to a Successful Military-to-Civilian Transaction
©2012, PotomacBooksInc.com

DEDICATION

Dedicated to all active, retired, and former United States military and civilian law enforcement personnel. Thank you for your service.

PREFACE

As a longtime collector of U.S. Navy uniforms and rating badges, it seems natural that I would obtain a few other pieces of USN memorabilia over the years.

Over the last 25 years of collecting U.S. Navy memorabilia, I had obtained six or seven different Navy master-at-arms badges.

About three years ago, I attended the Tennessee Military Collectors Association annual show in Nashville. At his booth, a vendor displayed a small collection of Navy police badges. All those badges were for sale, and I purchased about half the 12 badges he displayed.

These badges included World War II-era Navy Police Petty Officer (PPO) badges. I had read copies of the *Blue Jacket's Manual* in the early days of my Navy career and was familiar with the term "police petty officer." I even remembered the duties of the PPO. None of the five ships I was attached to had ever bestowed the term police petty officer upon junior petty officers in any of the divisions to which I was assigned. The duties I understood as belonging to a police petty officer were instead assigned to the "duty PO" in the divisions to which I was assigned.

After that Nashville show, I became much more interested in law enforcement badge collectors and collections. I now have more than 100 examples in my collection, and I have another 300 images of scarce and rare USN law enforcement-style badges.

Jeweler-made badges, especially "presentation" badges, are not uncommon in civilian law enforcement (LE) circles but exceptionally rare in military law enforcement badge collections. Because almost every Navy LE badge was procured and paid for via the USN supply chain, no exorbitant prices would have been authorized. I consider myself lucky to have in my collection two quite rare USN master-at-arms badges containing jeweler's marks (see pages 25 and 33.)

I was fortunate to find a few law enforcement collectors and an ex-sailor collector who have collected these items for many years. These individuals freely shared their accumulated knowledge with me. Therefore, I am indebted to them.

ACKNOWLEDGEMENTS

The author thanks Betty Thomas Smith and Angela Smith Ballard for their professional editing assistance and Daniel D. Smith, Jr. for his first-class graphic arts services.

Thanks also to the United States Militaria Forum members David W. Fox and Joe Saitta for their images, technical input, and draft review efforts.

Any errors are wholly mine, and corrections are welcomed.

Yokosuka, Japan 1969

CONTENTS

Rel. 2.1

Daniel D. Smith, Sr.

U.S. Navy Law Enforcement-Style Badges

History of Municipal Police Badges in the United States

U.S. Navy Law Enforcement (LE) badges are a sub-set of American law enforcement badges. For that reason, let's look first at the historical context of municipal police badges in the USA.

Dictionaries define *the badge* as a mark, token or device worn as a sign of allegiance, membership, authority and achievement.[i]

The history of the first use of badges to denote peacekeeping authority has multiple interpretations of "*the beginnings.*"

Some of the confusion surrounding the beginnings can be contributed to your definition of peace keeping, or law and order authority. For example, would *King Arthur's Knights of the Round Table* fit this description? Some collectors believe King Arthur's knights' shields were the first symbols of authority.

Others believe that "*from the livery badge, various badges of service evolved, worn by officials, soldiers and servants. In the British army a metal (today often plastic) cap badge denoting the soldier's regiment became standard by the 17th century, as in most European armies (though not always navies). By the 19th century a badge was an almost invariable part of any uniform, including school uniforms, which in the UK usually still feature the school's badge in cloth on the breast pocket of the jacket or blazer.*"[ii]

An excellent definition comes from Timothy Roufa, who posted, [P]*erhaps the most visible and recognizable symbol of policing around the world, the police badge is seen by many as a sign of authority, sacrifice and service. While most closely associated with public service professionals – fire fighters, correctional officers and, especially, police officers – the use of badges actually long predates the modern concept of law enforcement and policing as we know it.*

When thinking of the word 'badge,' one most likely pictures the metal shield, star or similar emblem worn by officers and firefighters. In fact, the term is used to describe any smaller object – such as a name tag, pin, emblem or other device – that can be displayed, often prominently, to show who a person is. This is the concept from which the badge originated and from where the modern police badge evolved.

European nobility are known for their use of heraldry, coats of arms and insignia to represent their House and history. During the Middle Ages, a time when power was not consolidated and national borders were far more fluid than in modern times, badges were worn by commoners and nobility alike to show their allegiance to a House or a group, and the wearing [of] *badges was not limited to a specific job or duty.*

These badges were made of various materials, including cloth, metal and even jewels.

Eventually, servants and others tied through employment or oaths to Houses often wore livery – clothing, colors and emblems – to display who they worked for and to identify their status. Depending on the relationship or level of service to the House's Seat, badges were issued and worn. Over time, the use of badges became more and more restricted to House servants, especially those with authority.

From these livery badges, badges of service ultimately developed, and thus public servants became easily identifiable by the prominent display of their badge. It was only natural, than [sic]*, that when the modern police force evolved, officers would adopt the badge to indicate their position, authority and service.*

As the profession of policing has evolved, its most visible symbol of service has taken on an almost iconic meaning. To the police officer, the badge represents the public trust, with which she has the authority to act and to which she has the duty to remain true.

From a practical standpoint, officers wear badges as a means to identify who they are and who they work for. The law enforcement badge is an easily recognizable sign of legal authority. From an altruistic perspective, though, uniformed police officers proudly wear their badges to show their allegiance to the communities they serve and pride they have in their chosen profession.[iii]

Since the United States of America is a relatively young nation, we have better documentation of our municipal history than most Old-World countries. The usage of the badge in our largest municipalities is fairly well documented.

The first official police force in the USA is attributed to the city of Boston, Massachusetts. The second was New Amsterdam (New York City), and the third was Philadelphia, Pennsylvania.

Philadelphia claims first use of a metal badge for law enforcement (May 1845), then New York City (July 1845), Baltimore, MD, (October 1851) and Boston, MA, (April 1853).[iv]

Abbreviated United States Municipal Police Timeline

Rattle Watch (1700s), (mainly a fire watch); Commissioned Police Officers (1850); U.S. Marshals (1870), who sometimes created a badge from a used baking soda can, resulting in the nickname "tin star," also created badges from silver coins, especially Mexican silver coins; Roundsman (1900) and Patrolman (1920).

Navy Police Badges

After the disestablishment of the U.S. Navy's Master-at-Arms (MAA) rating specialty in 1921, Masters-at-Arms were only distinguishable from other enlisted crew members by their wearing of a brassard, or arm band, with the letters "MAA" on the sleeve of their working uniforms.

Between 1925 and 1941, U.S. Navy commands, both ship and shore, locally purchased and issued metal badges to Masters-at-Arms and Police Petty Officers. These badges were similar to civilian law enforcement agencies. That ordering process, however, caused for badge inconsistencies throughout the Navy in terms of the size, color, materials and inscriptions when compared to the uniformity of the other service branches' military police forces.

U.S. Navy Master-At-Arms

In the United States Navy, the Master-at-Arms rating specialty, sometimes referred to as the Master-at-Arms "force," is the main equivalent of a police officer as most understand the term "police officer."

The Master-at-Arms (MA) rating was one of the original Navy ratings when it was first established in 1797. The MA rating was disestablished in 1921, and the responsibilities of the Master-at-Arms rating were assigned as collateral duties to other ratings.

In the Navy regulations of 1833, "marks of distinction" were prescribed for Petty Officers and other enlisted sailors for their "outside dress." This would be the first prescribed uniforms and insigne for enlisted men of the U.S. Navy. The insigne were described thus:

*"Petty Officers...The anchor shall be not more than three, nor less than two inches in length, placed half way between the elbow and shoulder, upon the front of the sleeve: it shall be white when worn on a blue garment, and blue when worn on a white garment."*v

Sewing template for 1833 USN "Mark" of Distinction (Image: 1)

In 1841 the Mark was changed to represent an eagle and anchor

1841 eagle and anchor for USN Petty Officers (sewing stencil for white uniforms) (Image: 2)

In 1866 eight "marks" were introduced to be worn with the updated Petty Officer mark. These new marks were *to distinguish the job function of the wearer.* These were the first of what would come to be called "specialty" marks. The original specialties were Master-at-Arms, Gunner's Mates, Coxswains, Boatswain's Mates, Quartermasters, Carpenter's

Mates, Captain of the Tops and Sailmaker's Mate. The marks were worn on the right or left sleeve depending on which "watch" section the sailor belonged to, either the Port (left) or the Starboard (right) watch section.

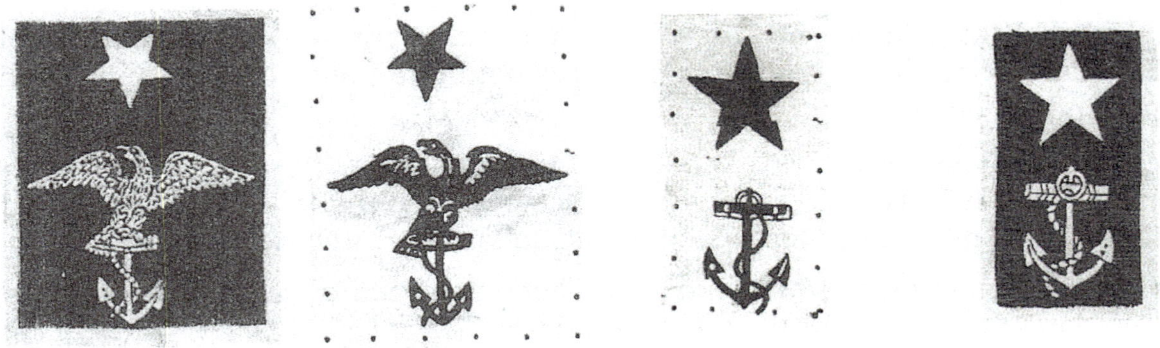

A. Petty Officer's Marks

B. Master at Arms

Two of the "Marks" introduced in 1866 (shown are sewing templates) (Image: 3)

Photo # NH 56547 USS Boston's Master-at-Arms mess, 1888

Pre-metal badge days (Image: 4)

(Image 5)

The above photo from 1925 shows two sailors with typical duty gear of web belt, baton and leggings. The far sailor has a brassard on his arm and is wearing a badge; the brassard is too dark to make out the details. The closer sailor is carrying three brooms and appears to be on the way to pick up a work detail. Until 1921, when the Master-at-Arms rating was disestablished, these sailors would likely have been wearing Master-at-Arms rating badges.

The photo has "Ref. Dept., MAR 20, 1925, ...taken on our muny pier..." on the reverse. The lady who wrote on the back of the photograph, "Deb," refers to the sailors as "naval MP." Since the Army instituted MPs in their reserve program in 1921, it is possible the term "military police" was more familiar to the civilian population at that point in time.

USN Master-at-Arms (1886-1893) (Image: 6) USN CMAA (1893-1921)

Reestablished on August 1, 1973, the Master-at-Arms (MA) rating would initially only draw its members from sailors cross-rating from other active ratings at the second class petty officer level and above. This process continued until 2003, when the Navy Recruiting Command was assigned a new contract mission for entry-level applicants into the MA specialty field.[vi]

Standardization of Badge Usage

No initial directive on which particular badges to order left all units, ship- and shore-based alike, to order, generally, from the nearest law enforcement-style badge vendor. Some units went with the largest badge vendors at the time; some units, having large repair facilities aboard their ship such as repair tenders, had a badge produced by crew members. These conditions resulted in many different styles, sizes, materials and descriptive wording being used on the earliest Navy Master-at-Arms (MAA) and Police Petty Officer (PPO) badges. Shore Patrol (SP) badges did not appear until after July 1944. Current Master-at-Arms badge usage is dictated by Navy uniform regulation NAVADMIN 164/12, dated Apr. 2012, and referencing 164/12 dated 01JAN03.

The standard badge size stated is 2.50"(h) x 1.75"(w). The earlier "Chief Master-at-Arms," "Assistant Chief Master-at-Arms," and "Master-at-Arms" designations are now standardized to "Command Master-at-Arms (CMAA)."

The following verbiage is from the official U.S. Navy directive concerning uniform regulations dictating the use of USN badges by security forces, corrections specialists, and masters-at-arms.

SUBJ: UNIFORM BOARD UPDATE
NAVADMIN 164/12
REF/A/DOC/NAVPERS 15665I/*01JAN03*//
REF/B/MSG/CNO WASHINGTON DC/301439ZAUG11//
REF/C/MSG/CNO WASHINGTON DC/022305ZDEC11//
REF/D/MSG/CNO WASHINGTON DC/261511ZJUN09//

NARR/*REF A IS U.S. NAVY UNIFORM REGULATIONS. REF B IS NAVADMIN 259/11.*
REF C IS NAVADMIN 366/11. REF D IS NAVADMIN 186/09.
RMKS/1. THIS NAVADMIN ANNOUNCES THE CHIEF OF NAVAL OPERATIONS (CNO) APPROVED UNIFORM POLICY CHANGES PROPOSED DURING THE 24 JANUARY 2012 AND 3 APRIL 2012 NAVY UNIFORM BOARDS. THE NAVY UNIFORM BOARD IS LED BY THE CHIEF OF NAVAL PERSONNEL (PRESIDENT) AND INCLUDES REPRESENTATIVES FROM THE FLEET, OPNAV, RESERVE FORCES, COMMANDER, NAVAL INSTALLATIONS COMMAND, NAVAL SUPPLY SYSTEM COMMAND, NAVAL EDUCATION AND TRAINING COMMAND, AND U.S. NAVAL ACADEMY.
2. APPROVED UNIFORM POLICY CHANGES ARE AS FOLLOWS:

A. ***NAVY SECURITY FORCE IDENTIFICATION BADGES. U.S. NAVY SECURITY FORCE IDENTIFICATION BADGE/PATCH WILL BE STANDARDIZED IN DESIGN AND COLOR***

REFLECTING THREE CATEGORIES OF SECURITY DUTIES:
U.S. NAVY SECURITY FORCES, U.S. NAVY CORRECTIONS SPECIALIST AND U.S. NAVY COMMAND MASTER-AT-ARMS.

(1) DESCRIPTION: SECURITY BADGES WILL BE METAL OR EMBROIDERED PATCHES, GOLD IN COLOR, BLACK LETTERING AND APPROXIMATELY 2 1/2 INCH IN HEIGHT BY 1 3/4 INCH IN WIDTH. METAL BADGES WILL BE STAMPED WITH RAISED INSIGNIA. THE FACE OF THE BADGE/PATCH WILL HAVE A RIGHT FACING EAGLE PERCHED ATOP A WREATHED SHIELD. ADDITIONALLY...

...(C) COMMAND MASTER-AT-ARMS (CMAA): THE TOP SCROLL (LINE 1) WILL BE STAMPED U.S. NAVY; THE SECOND SCROLL (LINE 2) WILL BE STAMPED CMAA, THE THIRD SCROLL (LINE 3) WILL BE BLANK; AND THE FOURTH SCROLL (LINE 4) WILL HAVE A BLACK EMBOSSED STAR. THE CENTER PIECE WILL BE A FOULED ANCHOR.

Master-at-Arms Petty Officer First Class 1973, current CMAA badge
(Image: 7)

From 1921, when the Master-At-Arms (MAA) specialty was discontinued, through the late 1930s, USN personnel recognized their unit's Master-at-Arms by the presence of an MAA brassard, or arm band, worn on their working uniforms.

MAA brassard (stylized) (Image 8)

Rarity of the Badges Covered in This Book

Many aspects of a badge must be considered when rating the rarity of a collectible badge. Some of these considerations would be the number of badges produced, the condition, the age, presence of maker hallmarks, and whether the badge was produced by a large manufacturer, custom-made by a jeweler or locally handmade.

Also considered are the materials used in the production of the badge as well as the size of the military unit that used the badge. Is the badge marked sterling or some other precious metal? Are jeweler hallmarks present? While it is subjective, eye appeal matters. Does the badge "grab" your attention? Provenance is important in authenticating the authenticity of a badge.

Famous units such as the large battleships and famous aircraft carriers often result in more desirability. Those who make fake badges know that famous unit badges are more desirable, so buyers should be even more cautious when buying badges from famous units such as a USS Midway, a battleship, or a Pearl Harbor badge. On the other hand, small units, such as Patrol Torpedo Boats (PT boats) have only a handful of personnel assigned, and the number of badges issued to such a unit is very small, often only one. That results in greater difficulty in obtaining a badge from a small unit.

Considering these aspects of a badge, the following is a rarity rating scale for the badges found in this work. Not everyone will agree with the rarity values assigned to the badges in this work, and all opinions are respected.

Scale	*Description*
A	**Common**
B	**Not so common**
C	**Scarce**
D	**Rare**
E	**Very Rare**

Basic Badge Shapes and Styles

Various badge experts, such as Dave Arnold (BadgeCollecting.com), Dale Fenton (BadgeValues.com), and James C. Casey (Badges of America's Finest), have categorized badge styles using various law enforcement badge manufacturers' catalogs.

These collecting experts have often used a *minimalist* approach in classifying badge shapes (e.g., star, shield, oval, round).

Other collectors have presented a more detailed definition of badge shapes, most of which are illustrated below.

SHIELDS

Pinched Shield

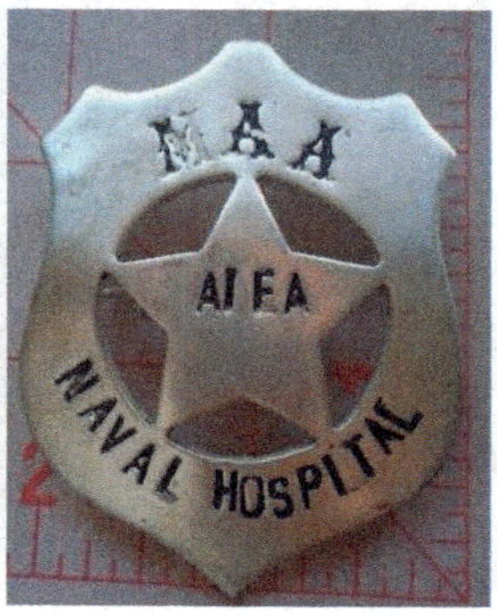

Pinched Shield w/Cut-out Star

Eagle-Top Shield (a.k.a. Teardrop)

Shield w/Sunburst

Image (9)

SHIELDS

Flat Sided Shield

Sunburst Shield

Eagle-Top Sunburst

Eagle-Top Shield

Image (10)

SHIELDS			

Eagle-top Shield	Shield	Eagle-Top Shield	Eagle-Top crossed baton hat badge

Image (11)

STARS (Stars are also named by the number of points, five-point, seven-point, etc.)			
	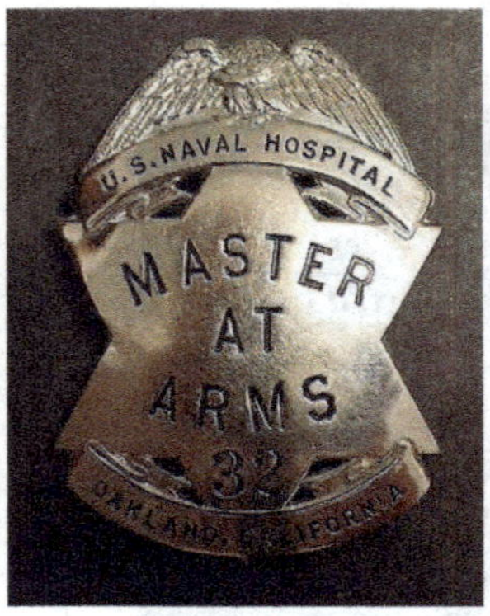		
Custom Ball-Tipped Star	6-Point Eagle-Top Star	6-Point Flat Ball-Tipped Star	6-Point Round Ball-Tipped Star w/banners

Image (12)

STARS			

Circle-Star Cutout	5-Point Round Ball-Tipped Star	7-Point Star	6-Point Star

Image (13)

<table>
<tr><th colspan="4">Ovals-Buttons-Rounds</th></tr>
<tr>
<td>
</td>
<td>
</td>
<td>
</td>
<td>
</td>
</tr>
<tr><td>Button (a.k.a Round)</td><td>Celluloid Other</td><td>Oval</td><td>Celluloid Round</td></tr>
</table>

Image (14)

Earliest Master-at-Arms Metal Badge Usage

The disestablishment of the Navy's Master-at-Arms rating in 1921, as well as the establishment of the Army Reserve program's Military Police Corps in that same timeframe, had some impact on the beginning of the U.S. Navy's use of law enforcement-style badges.

The earliest usage of a metal badge by Masters-at-Arms in the USN is dated to the late 1920s to mid-1930s. The oldest USN-related law enforcement-style badge found by this author is a 1920s contract police officer badge from the Aeronautic Station at Pensacola, Florida. The Navy changed the terminology of its aircraft program from Aeronautic Stations to Naval Air Stations in the late 1920s. Pensacola, Florida, was the U.S. Navy's first Naval Air Station and is home to the Naval Air Museum.

The USS Chamount (AP-5), a troop carrier, was turned over to the U.S. Navy in November 1921. She was home-ported in San Francisco, California.

The frequent onboard presence of hundreds of soldiers, marines and Construction Battalion (CB) personnel might have contributed to the ship's company feeling the need for a Master-at-Arms force that was closer to a civilian police presence than most Navy commands deemed necessary. On other Navy ships, the crew members likely knew all the MAA personnel aboard their ship, and the usage of an MAA brassard on the arm of the working dungaree uniform was presence enough for those MAA forces.

For whatever reason, the USS Chaumont was, if not the first, one of the very first U.S. Navy ships to utilize a law enforcement-style badge for its onboard MAA force. The majority of MAA and PPO badges did not enter widespread usage until the massive influx of wartime USN personnel beginning in early 1942.

Additionally, the badge, (Image 15), is the only badge this author has come across that utilizes the term "Deputy" MAA. The term was much more prevalent in civilian law enforcement. Later USN regulations would call for the appointment of an MAA and an Assistant Master-at-Arms.

Image 15
Shape: Shield with stylized anchor
Const/Rarity: Custom Die-cut / E
Clasp style: Pocket clip
Item Notes: Believed to be one of the earlier MAA badge usages. Possibly jeweler-made in the mid-1920s or early 1930s.
Images: FlyingTigerAntiques.com

Master-at-Arms Afloat

Image 16	Image 17
Shape/Size: Eagle-top shield, 3.2"x2.25"	Shape/Size: Eagle-top teardrop, 3.3"x2.5"
Const/Rarity: Die-cut / C	Const/Rarity: Die-cut, heavy badge / B
Clasp style: Wire pin and fork catch	Clasp style: Wire pin and small Burgess catch
Item Notes: Scarce "Afloat" Master-at-Arms	Item Notes: Stamped G93 on reverse; #23 seems high; applied crest and scrolls result in a multi-level badge
Author's collection	Author's collection

Image 18	Image 19
Shape: Eagle-top shield	Shape: Eagle-top teardrop shield
Const/Rarity: Die-cut / B	Const/Rarity: Die-cut / B
Clasp style: Bar pin catch; note solder holes for crest [or seal] attachment	Clasp style: Wire pin and fork catch
Image: FlyingTigerAntiques.com	Image:

Image 20	Image 21
Shape/Size: 5-point Star w/flat ball tips, 1.5"x1.5"	Shape/Size: Eagle-top shield, 3"x2.25"
Const/Rarity: Die-cut / B	Const/Rarity: Die-cut / B
Clasp style: Wire pin and small Burgess catch	Clasp style: Wire pin and tongue catch
Item Notes: HM Goldmark, Norfolk, VA E.D.F. (enlisted dining facility), very small badge	Item Notes: No reverse die. #3 applied, application of numbers and crests makes the badge a multi-level badge
Author's collection	Author's collection

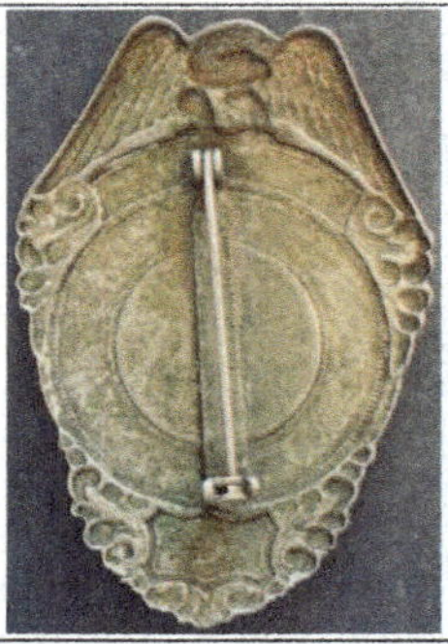

Image 22	Image 23
Shape/Size: Eagle-top shield, 2.85"x 2"	Shape/Size: Eagle-top shield, 2.30"x1.50"
Const/Rarity: Die-cut / C	Const/Rarity: Die-cut / C
Clasp style: Bar pin	Clasp style: Pin and large Burgess catch
Item Notes: 1955-58 usage, uncommon crest	Item Notes: 1966-93 hard enamel lettering
Author's collection	Author's collection

Image 24	Image 25
Shape: Eagle-top shield	Shape/Size: Eagle-top shield, 2.50”x1.70”
Const/Rarity: Die-cut / B	Const/Rarity: Die-cut / B
Clasp style: Bar pin	Clasp style: Pin and small Burgess catch
Item Notes: Considered a “catalog badge,” this eagle-top shield design can be traced to civilian use as early as 1875.[vii]	Item Notes: Addition of crest after initial die cut makes this a two-layer badge; the majority of crest badges are multi-layer.
Images:	Author’s collection

Image 26	Image 27
Shape: Eagle-top shield	Shape/Size: Eagle-top shield, 2.3”x1.4”
Const/Rarity: Die-cut / B	Const/Rarity: Die-cut / B
Clasp style: Pin and small Burgess catch	Clasp style: Pin and broken catch
Item Notes: Combination hard and soft enamel lettering, scarce crest design	Item Notes: USS White River, Korean War usage
Author’s collection	Author’s collection

Image 28	Image 29
Shape/Size: Eagle-top shield, 2.85”x2.25”	Shape: Sunburst-top shield
Const/Rarity: Die-cut / B	Const/Rarity: Die-cut / C
Clasp style: Bar pin	Clasp style: Pin and tongue catch
Item Notes: The green tint (verdigris) is from long contact with a leather pocket hanger	Item Notes: Hard enamel lettering
Author’s collection	Image:

Image 30	Image 31
Shape/Size: 6-pt Ball-tipped star, 2.5"x2.25"	Shape: Eagle-top shield
Const/Rarity: Die-cut / C	Const/Rarity: Die-cut / B
Clasp style: Bar pin	Clasp style: Bar pin
Item Notes: Unique lettering, die-cut ball tips	Item Notes: 1976-86
Author's collection	Images:

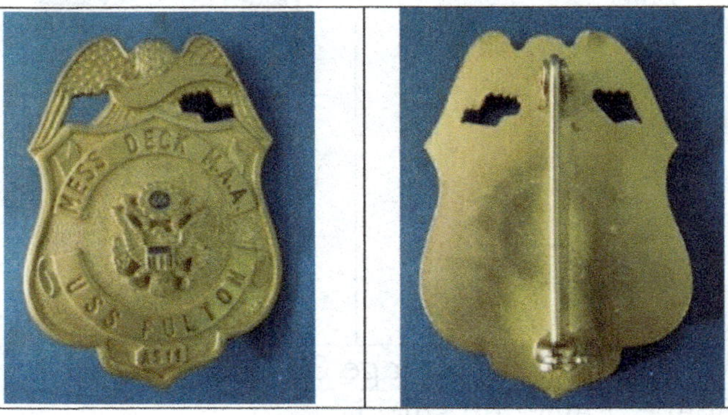	
Image 32	Image 33
Shape/Size: Eagle-top shield, 2.55"x1.85"	Shape: Eagle-top teardrop shield
Const/Rarity: Die-cut / B	Const/Rarity: Die-cut / A
Clasp style: Pin and large Burgess catch	Clasp style: Pin and small Burgess catch
Item Note: Submarine tender Mess Deck MAA	Item Notes: Hard enamel; hallmarked
Author's collection	Images:

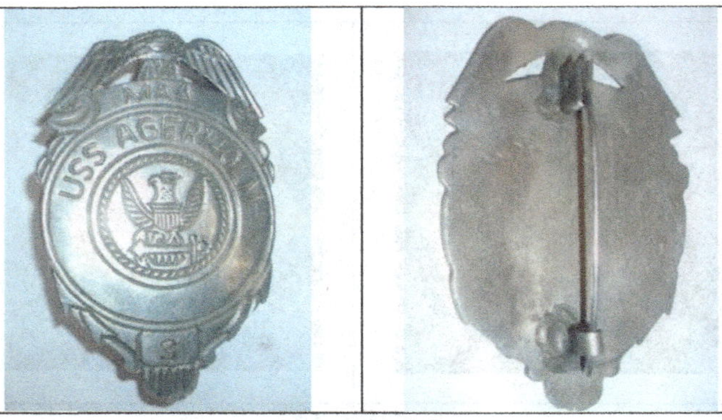	
Image 34	Image 35
Shape: Custom eagle-top shield	Shape: 7-point star
Const/Rarity: Die-cut, custom engraving / C	Const/Rarity: Die-cut / B
Clasp style: Pin and tongue catch	Clasp style: Pin and ball catch
Item Notes: Possibly foreign made	Item Notes: Soft enamel lettering
Images:	Images:

Image 36	Image 37
Shape: 6-point, Ball-tipped star	Shape: Eagle-top shield
Const/Rarity: Die cut / B	Const/Rarity: Die-cut / B
Clasp style: Pin and fork catch	Clasp style: Pocket clip
Item Notes: Hallmarked; soft enamel lettering	Item Notes: Believed to be 1943-46 usage
Images:	Images:

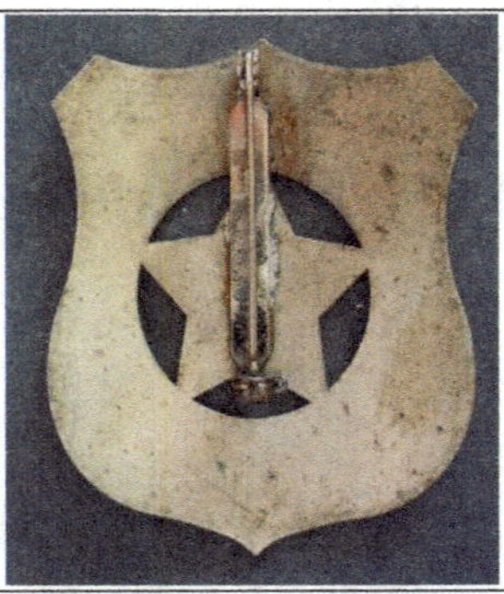

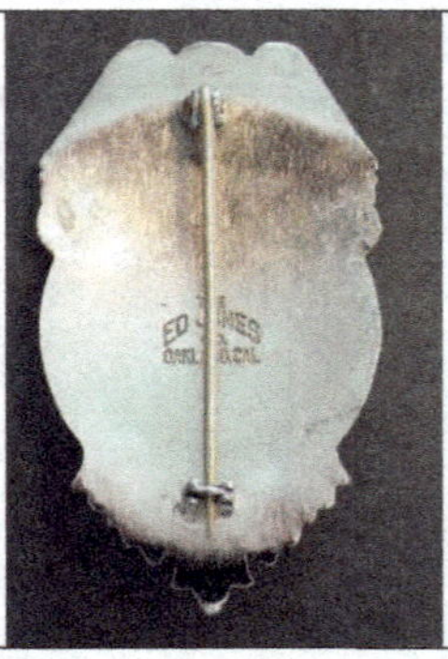

Image 38	Image 39
Shape: Shield w/cut-out star	Shape: Eagle-top shield
Const/Rarity: Die-cut / B	Const/Rarity: Die-cut / C
Clasp style: Bar pin	Clasp style: Pin and fork catch
Item Notes: Originally an inexpensive badge sometimes referred to as a “catalog badge” or “stock badge.” Note different MAA engraving	Item Notes: HM Ed Jones, “Ship’s” MAA while commonly used aboard ship, it is uncommon usage on badge itself.
Images:	Images:

Image 40	Image 41
Shape/Size: Eagle-top shield, 2.6”x1.65”	Shape: Sunburst shield
Const/Rarity: Die-cut / B	Const/Rarity: Die-cut / B
Clasp style: Pin and ball catch	Clasp style: Pin and Burgess catch
Item Notes: USS Coral Sea, 1952-90 usage	Item Notes: Gold tone badge, 1968-75 usage
Author’s collection	Image:

Image 42	Image 43
Shape: Eagle-top shield	Shape/Size: Sunburst-top shield, 2.85"x2"
Const/Rarity: Die-cut / B	Const/Rarity: Die-cut / C
Clasp style: Bar pin	Clasp style: Pin and small Burgess catch
Item Notes: Soft enamel lettering, 1963-79	Item Notes: Soft enamel lettering, 1950s-60s
Images:	Author's collection

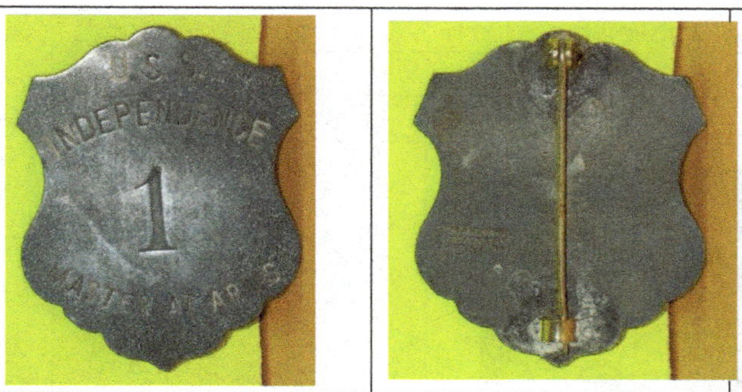

Image 44	Image 45
Shape: Pinched Shield	Shape/Size: E-T sunburst, 3.15"x2.30"
Const/Rarity: Die-cut / A	Const/Rarity: Die-cut / B
Clasp style: Pin and tongue catch	Clasp style: Pin and small Burgess catch
Item Notes: Hallmarked Philadelphia, PA	Item Notes: Soft enamel lettering, 1960-75
Image:	Author's collection

Image 46	Image 47
Shape: Eagle-top shield	Shape/Size: Eagle-top shield / 2.8"x 1.8"
Const/Rarity: Die-cut / B	Const/Rarity: Die-cut / B
Clasp style: Spring pin w/saddle	Clasp style: Pin and small Burgess clasp
Item Notes: Old style, 1956-60s usage	Item Notes: Note rear solder holes to secure crest
Images:	Author's collection

 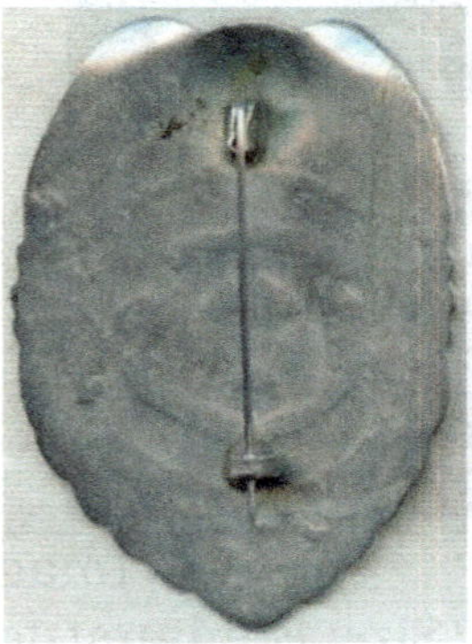	 
Image 48	Image 49
Shape/Size: Eagle-top shield, 2.7”x1.9”	Shape/Size: Eagle-top shield, 2.5”x1.75”
Const/Rarity: Die-cut / C	Const/Rarity: Die-cut / A
Clasp style: Wire pin and tongue catch	Clasp style: Pin and large Burgess catch
Item Notes: Likely made in Jan. 1969, Subic Bay, P.I. port-of-call.	Item Notes: 1990s-2009 usage, crest was part of die-cut pattern
Author’s collection	Author’s collection

Image 50	Image 51
Shape: Star	Shape: Eagle-top shield
Const/Rarity: Die-cut / B	Const/Rarity: Die-cut / C
Clasp style: Soldered safety-pin	Clasp style:
Item Notes: USCG crew on USN ship WWII, hand-lettered.	Item Notes: The Sculpin was commissioned in 1938; this badge may have been an early usage by units and since the MAA specialty ‘went away’ in 1921, this badge might have been ordered by personnel who confused civilian terminology with the position an MAA would hold aboard ship. Or this could be an “error” badge.
Images: The Wharfmaster collection	Image:

Image 52
Shape: Eagle-top shield
Const/Rarity: Die-cut / B
Clasp style: Pin and ball catch
Item Notes:
Images:

Image 53
Shape/Size: Eagle-top shield, 2.5"x1.45"
Const/Rarity: Die-cut / B
Clasp style: Bar pin
Item Notes: From the author's second ship assignment
Author's collection

Image 54
Shape/Size: Eagle-top shield, 2.5"1.7"
Const/Rarity: Die-cut / A
Clasp style: Clutch pins, 1980s
Item Notes:
Author's collection

Image 55
Shape/Size: E-T sunburst, 2.85"x2.30"
Const/Rarity: Die-cut / B
Clasp style: Pin and Burgess catch
Item Notes: Uncertain why fireman's crest
Author's collection

Image 56
Shape: Eagle-top shield
Const/Rarity: Die-cut / B
Clasp style: Pin and small Burgess catch
Item Notes: Ship's crest
Images: David Fox collection

Image 57
Shape/Size: Eagle-top shield, 3.4"x2.45"
Const/Rarity: Die-cut / B
Clasp style: Pin and large Burgess catch
Item Notes:
Author's collection

Master-At-Arms Ashore

Image 58
Shape/Size: Eagle-top shield, 3.35"x2.40"
Const/Rarity: Die-cut, 2-layer, stamped "Silver y.t.k." / E
Clasp style: Pocket clip
Item Notes: A very nice, rare jeweler or silversmith made early shore-based MAA design. The E.M. club, or enlisted man's club, located on most navy bases was where sailors ventured to drink alcohol on base. A Master-at-Arms presence would certainly seem to be called for when mixing alcohol and young sailors. Research turned up other pieces by this silversmith, but failed to identify the silversmith.
Images: Author's collection

MAA Force USS Courtney, 1963, Newport, R.I. (Image 59)

Image 60	Image 61
Shape/Size: Eagle-top shield, 3”x1.80”	Shape: Eagle-top shield
Const/Rarity: Die-cut / C	Const/Rarity: Die-cut / B
Clasp style: Wire pin and fork catch	Clasp style: Pin with Burgess catch
Item Notes: Early WWII usage	Item Notes: NAS, North Is., San Diego CA
Author’s collection	Images:

Image 62	Image 63
Shape/Size: Eagle-top shield, 3.15”x2.10”	Shape/Size: Eagle-top shield, 3”x2.35”
Const/Rarity: Die-cut / B	Const/Rarity: Die-cut / B
Clasp style: Wire pin and large Burgess catch	Clasp style: Wire pin w/saddle, sometimes called a spring pin with saddle
Item Notes:	
Author’s collection	Author’s collection

Image 64	Image 65
Shape/Size: 7-pt star, 2.4”x2.4”	Shape/Size: Shield, 1.75”x1.65”
Const/Rarity: die-cut / A\B	Const/Rarity: die-cut and engraved / C
Clasp style:Wire pin and small Burgess catch	Clasp style: Wire pin and Burgess catch
Item Notes: Slightly vaulted (arched) badge, sometimes called “dapped”	Item Notes: Scarce design
Author’s collection	Author’s collection

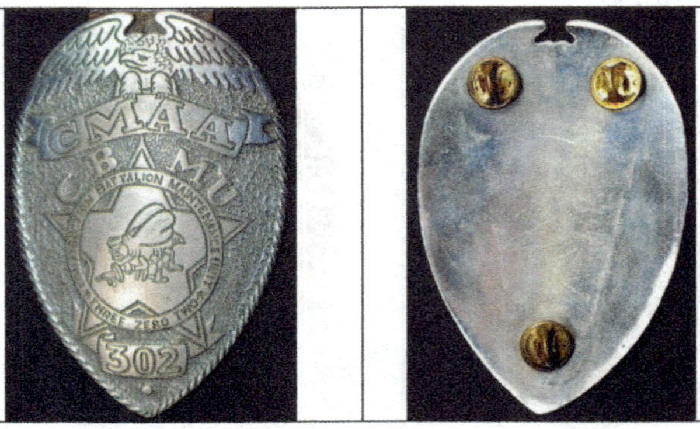

Image 66	Image 67
Shape/Size: Eagle-top shield / 3.10”x2.05”	Shape: Eagle-top shield
Const/Rarity: Die-cut / C	Const/Rarity: Die-cut / C
Clasp style: Clutch pins / pocket hanger	Clasp style: Bar pin
Item Notes: Custom made in the Philippine Islands in the early 1970s. As found, had been converted to pocket hanger style.	Item Notes: Note “Galley” MAA; NAF Weeksville, NC, was a lighter-than-air (blimp) facility from 1942-57.
Author’s collection	Images:

Image 68	Image 69
Shape/Size: Shield, 2.6”x2.2”	Shape: Eagle-top shield
Const/Rarity: Die-cut / B	Const/Rarity: Die-cut / B
Clasp style: Wire pin and large Burgess catch	Clasp style: Pin and tongue catch
Item Notes: Gold-tone badge	Note: Subic Bay, P.I., local made
Author’s collection	Images:

Image 70	Image 71
Shape/Size: Shield, 2.55”x2.15”	Shape/Size: Eagle-top shield, 2.45”x1.75”
Const/Rarity: Die-cut / C	Const/Rarity: Die-cut / B
Clasp style: Soldered safety pin	Clasp style: Wire pin and tongue catch
Item Notes: Custom engraved by hand, scarce	Item Notes: Custom made, Philippine Isl.
Author’s collection	Author’s collection

Image 72	
Shape/Size:	6-Pt, ball-tip star, 2.55"x2.2"
Const/Rarity:	Die-cut / B
Clasp style:	Bar pin
Item Notes:	WWII, TI, San Fran., CA, (HM)
Author's collection	

Image 73	
Shape:	Eagle-top shield
Const/Rarity:	Die-cut / C
Clasp style:	Pocket clip
Item Notes:	Antarctica "USNMAA," scarce command
Images:	

Image 74	
Shape/Size:	Shield, 3"x1.75"
Const/Rarity:	Molded plastic / D
Clasp style:	Single wire pin w/saddle
Item Notes:	Rare, plastic, Hawaii base use
Author's collection	

Image 75	
Shape/Size:	5-Pt Ball-tipped star, 2.75"x2.90"
Const/Rarity:	Die-cut / C
Clasp style:	Wire pin and tongue catch
Item Notes:	Custom made, Guam unit
Author's collection	

Image 76	
Shape/Size:	E-T Shield, 2.40"x1.80"
Const/Rarity:	Die-cut / B
Clasp style:	Pin and tongue catch
Item Notes:	Many collectors call the sides of this design a pinched shield
Author's collection	

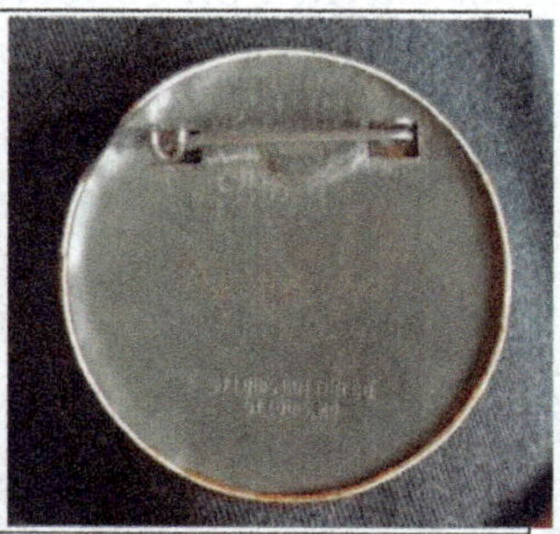

Image 77	
Shape/Size:	Round or Button, 2.20"x2.20"
Const/Rarity:	Laminated / C
Clasp style:	Single wire pin
Item Notes:	Quick, different, and inexpensive design
Author's collection	

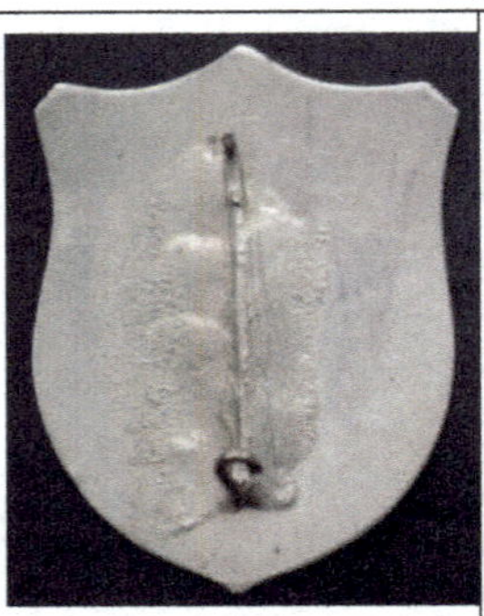

Image 78
Shape/Size: Shield, 2.50" x 2"
Const/Rarity: Die-cut / B
Clasp style: Single wire catch (broken)
Item Notes: Uncommon red lettering, PSNY name changed in 1945 to PSNS; "R.S." likely stands for Radio School
Author's collection

Image 79
Shape/Size: Sunburst w/eagle, 2.85"x2.35"
Const/Rarity: Die-cut / B
Clasp style: wire pin and small Burgess catch
Item Notes: Orlando, Fla. "boot camp," hallmarked
Author's collection

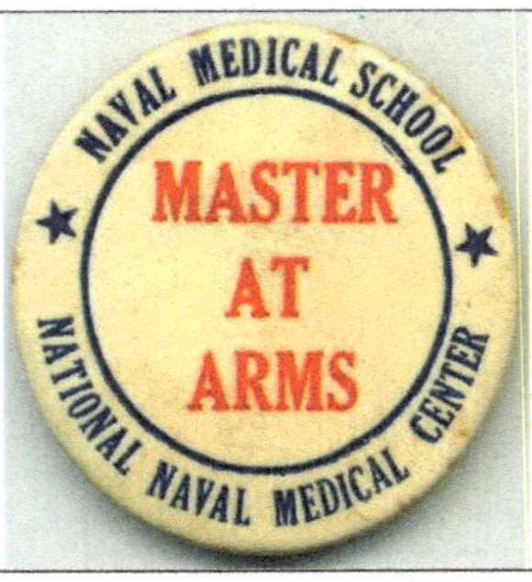

Image 80
Shape: Round or button
Const/Rarity: Laminated / B
Clasp style: Single wire pin
Item Notes: Hallmarked
Images: David Fox collection

Image 81
Shape: 6-pt Ball-tipped star
Const/Rarity: Die-cut / B
Clasp style: Pin and Burgess catch, either this pin is affixed upside down or the "2" is affixed upside down
Item Notes: Hallmarked
Images:

Image 82
Shape/Size: E-T Shield, 3.05"x2.40"
Const/Rarity: Die-cut / A
Clasp style: Bar pin
Item Notes: CB Battalion, HM

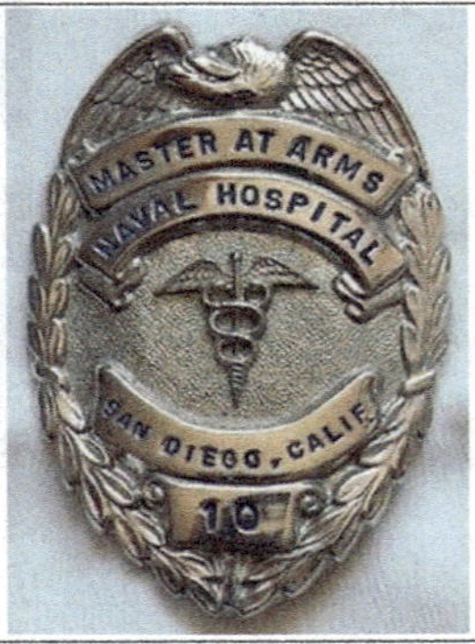

Image 83
Shape: Eagle-top shield
Const/Rarity: Die-cut / C
Clasp style: Wire pin w/small Burgess catch
Item Notes: Note Caduceus crest, scarce

Image 84	Image 85
Shape/Size: Eagle-top teardrop, 3.3"x2.3"	Shape: Eagle-top shield
Const/Rarity: Die-cut / B	Const/Rarity: Die-cut / C
Clasp style: Pin and large Burgess catch	Clasp style: Pocket clip
Item Notes:	Item Notes: Scarce unit, marked A'SST
Author's collection	Images:

Image 86	Image 87
Shape/Size: Eagle-top shield, 3"x2"	Shape/Size: Eagle-top shield, 2.4"x1.65"
Const/Rarity: Die-cut / B	Const/Rarity: Die-cut / B
Clasp style: Pin and large Burgess catch	Clasp style: Pin and ball catch
Item Notes: Resv. Cntr. Southfield, Mich.	Item Notes: HM
Author's collection	Author's collection

Image 88	Image 89
Shape: Eagle-top Shield	Shape/Size: Eagle-top shield, 2.3"x1.9"
Const/Rarity: Die-cut / C	Const/Rarity: Die-cut / C
Clasp style: Wire pin and ball catch	Clasp style: Spring pin w/saddle
Item Notes: Specific USN school listed	Item Notes: HM S G Adams Co
Images: David Fox collection	Author's collection

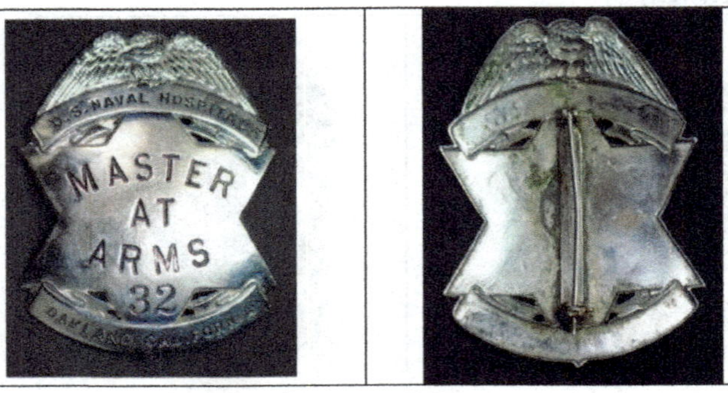

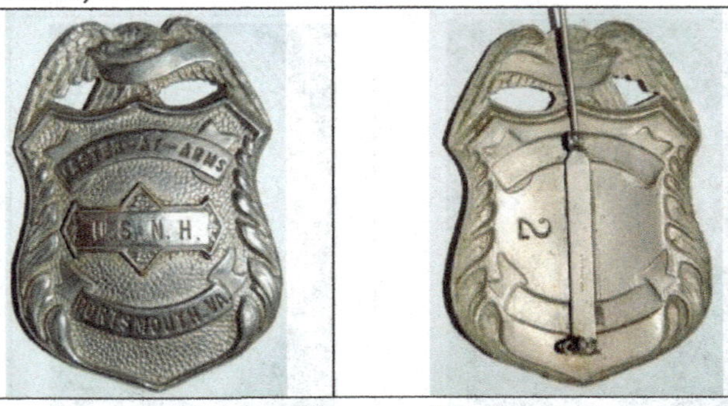

Image 90	Image 91
Shape/Size: E-T, 6Pt star, 2.95"x2.15"	Shape/Size: Eagle-top shield, 3.20"x2.2"
Const/Rarity: Die-cut / C	Const/Rarity: Die-cut / B
Clasp style: Bar pin	Clasp style: Bar pin
Item Notes: Uncommon Eagle-top star	Item Notes: Stamped "2" on reverse
Author's collection	Author's collection

Image 92	Image 93
Shape/Size: Shield, 2.35"x2.05"	Shape/Size: Sunburst w/eagle, 2.85"x2.35"
Const/Rarity: Die-cut / B	Const/Rarity: Die-cut / B
Clasp style: Spring pin w/saddle	Clasp style: Pin and large Burgess catch
Item Notes: Brass	Item Notes: State of Ill. crest, scarce
Author's collection	Author's collection

Image 94	Image 95
Shape: Shield	Shape/Size: 7-Pt star, 2.90"x2.80"
Const/Rarity: Die-cut / C	Const/Rarity: Die-cut / B
Clasp style: Bar pin	Clasp style: Bar pin
Item Notes: HM, U.S. NavSta Midway Island	Item Notes: Vaulted (stamped arch), if done manually, this would have been termed "dapping"
Images: David Fox collection	Author's collection

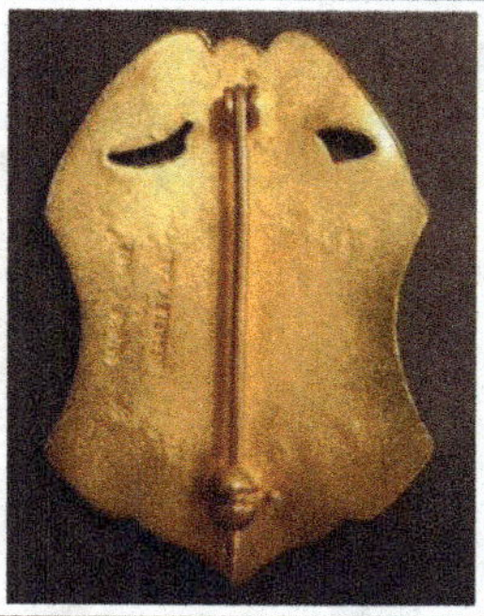

Image 96
Shape/Size: Eagle-top shield, 2"x1.4"
Const/Rarity: Die-cut / B
Clasp style: Pin and side ball-catch
Item Notes: HM George F. Cake

Image 97
Shape/Size: Eagle-top shield, 2.40"x1.7"
Const/Rarity: Die-cut / A
Clasp style: Spring pin
Item Notes: Originally an inexpensive and popular badge design.

Image 98
Shape: Pinched Shield
Const/Rarity: Die-cut / C
Clasp style: Single pin w/saddle
Item Notes: Hand stamped lettering, WWII
Images: David Fox collection

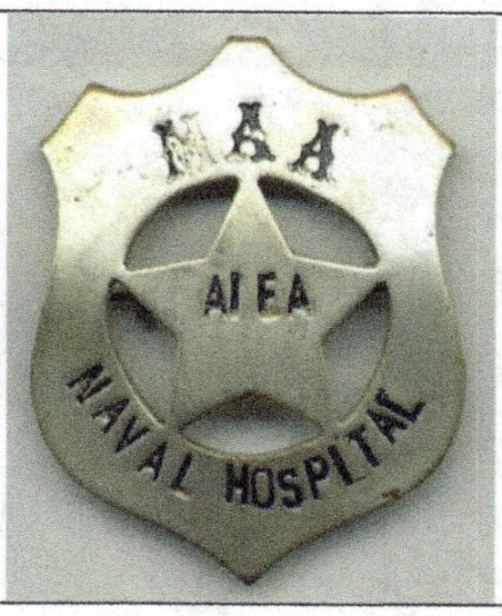

Image 99
Shape: Pinched Shield w/cut-out star
Const/Rarity: Die-cut / C
Clasp style: Bar pin
Item Notes: WWII, Territory of Hawaii.
Images: David Fox collection

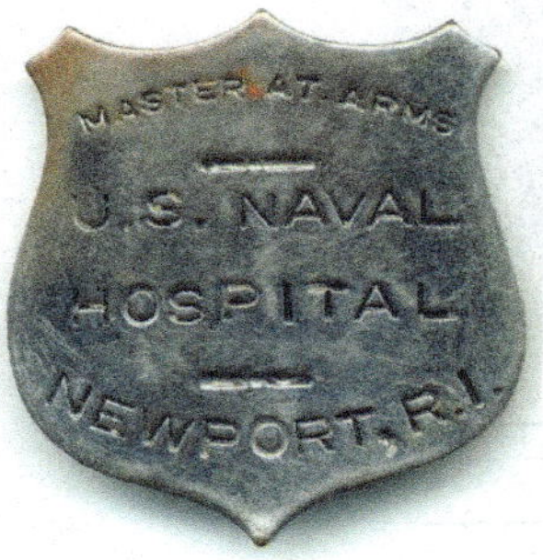

Image 100
Shape: Pinched Shield
Const/Rarity: Die-cut / B
Clasp style: Pin and fork catch
Item notes: Fork catch a.k.a. Fielding catch
Images: David Fox collection

Image 101
Shape: 6-point ball-tipped star
Const/Rarity: Die-cut / B
Clasp style: Spring pin w/saddle
Item Notes: P.E. School may be the 1940's diesel submarine and motor gunboat "Pancake" Engine School located in San Francisco, Calif., or it may be Primary Electronics School in WWII
Images: David Fox collection

Master-at-Arms (Generic / Unit Unknown)

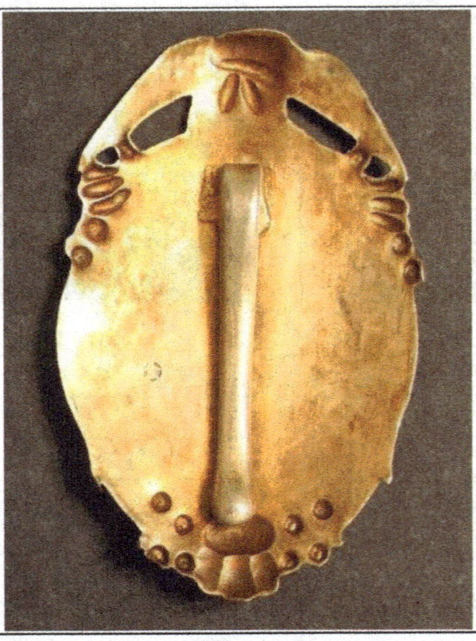
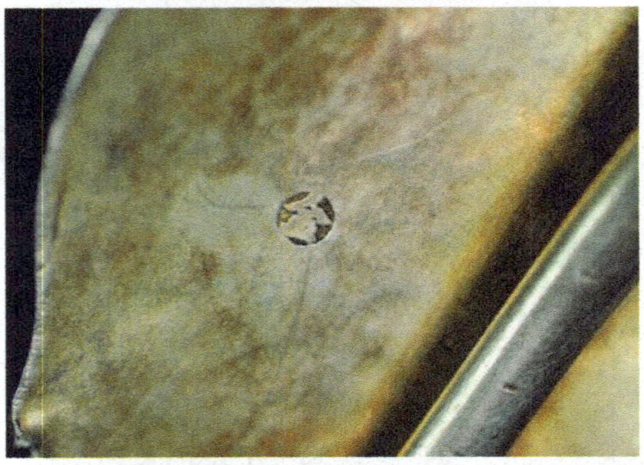

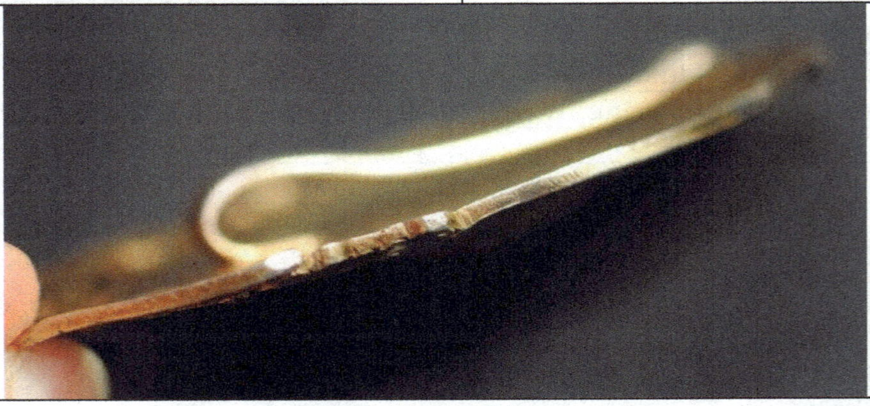

Image 102

Shape/Size: Eagle-top shield, 2.95"x1.90"

Const/Rarity: die-cut w/applied star / E

Clasp style: Pocket clip

Item Notes: A rare, jeweler-made MAA badge. Third-hand story has that this badge was owned by a past crew member of the USS Iowa. Don't know if it was worn on the Iowa. Mark on the reverse of the badge was that of the jeweler. Research failed to identify the jeweler.

Author's collection

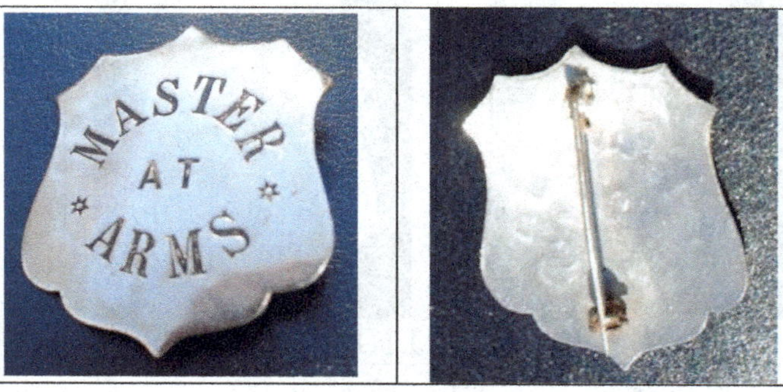

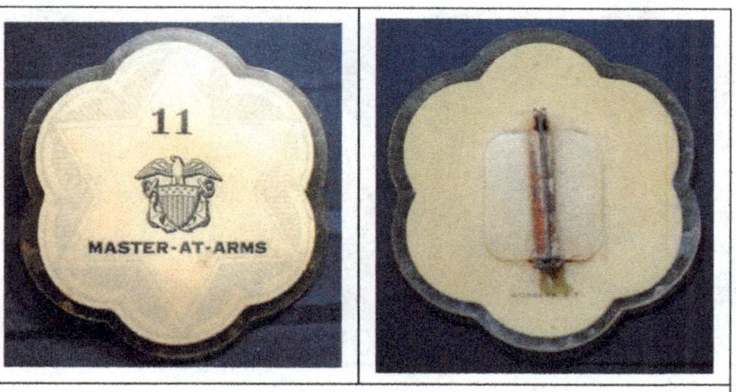

Image 103	Image 104
Shape/Size: Pinched Shield, 2.15"x1.80	Shape/Size: Shield, 3"x3"
Const/Rarity: Die-cut / A	Const/Rarity: Plastic, molded / C
Clasp style: Pin and small Burgess catch	Clasp style: Spring pin w/saddle
Item Notes: Simple pinched shield design	Item Notes: HM: Wobbers, S.F., a San Francisco Stationary and Engraving Bus.
Author's collection	Author's collection

Image 105	Image 106
Shape: 6-point, ball-tipped Star	Shape: Eagle-top shield
Const/Rarity: Die-cut / B	Const/Rarity: Die-cut / B
Clasp style: Spring pin	Clasp style: Pin and tongue catch
Item Notes: Good example of a "catalog" badge	Item Notes: Vintage
Author's collection	Images:

Image 107	Image 108
Shape: Eagle-top shield	Shape: Eagle-top shield
Const/Rarity: Die-cut / B	Const/Rarity: Die-cut / B
Clasp style: Bar pin	Clasp style: Pin and fork catch
Item Notes: Originally an inexpensive design	Item Notes: Marked "Sterling"; early
Images:	Images:

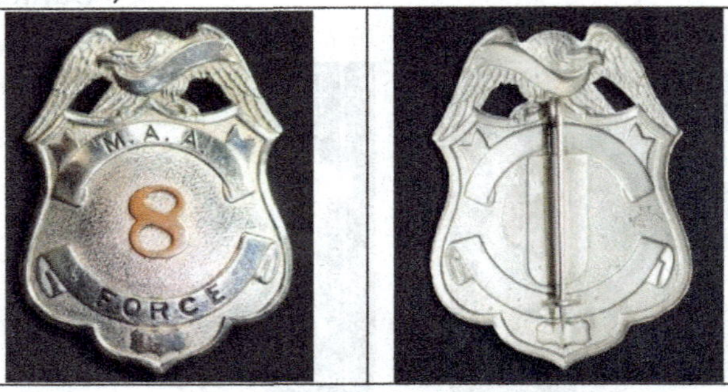

Image 109	Image 110
Shape/Size: Shield, 2.30"x2.25"	Shape/Size: E-T shield, 2.1"x1.6"
Const/Rarity: Die-cut / B	Const/Rarity: Die-cut / C
Clasp style: Pin and small Burgess catch	Clasp style: Spring pin w/saddle
Item Notes: Vintage	Item Notes: Nickle plating rubbed off "eight"; MAA "Force"; vintage usage
Author's collection	Author's collection

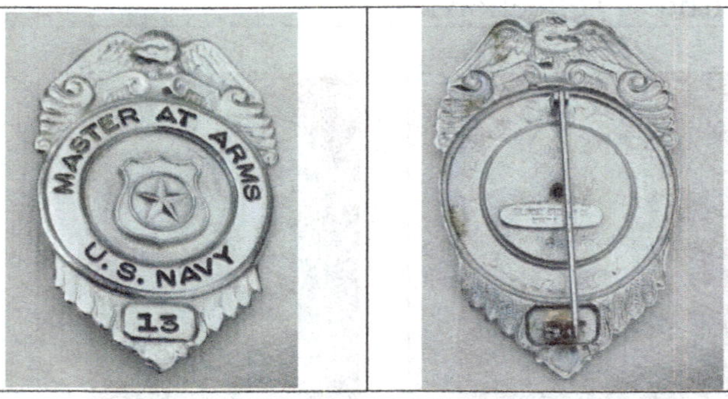

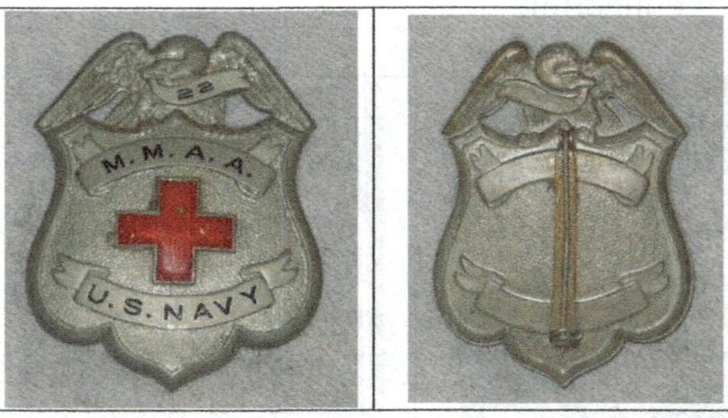

Image 111	Image 112
Shape/Size: Eagle-top shield, 2.9"x1.8"	Shape/Size: Eagle-top shield, 2.95"x2"
Const/Rarity: Die-cut / A	Const/Rarity: Die-cut / C
Clasp style: Pin and small Burgess catch	Clasp style: Bar pin
Item Notes: Hallmarked	Item Notes: Soft enamel lettering, Medical (dept.) MAA.
Author's collection	Author's collection

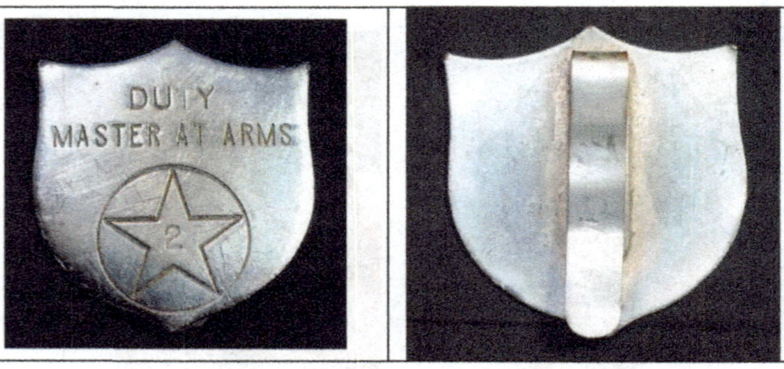

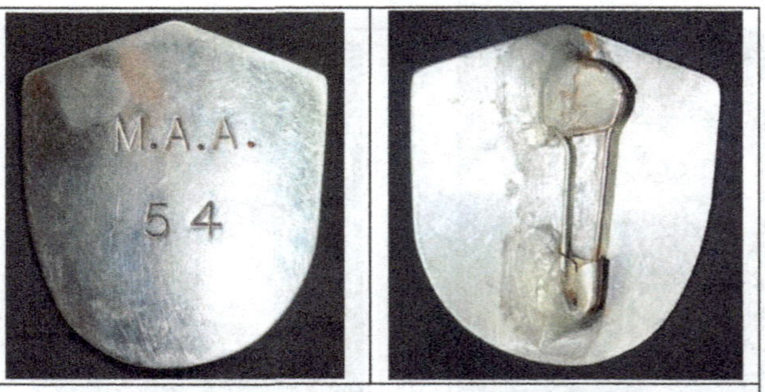

Image 113	Image 114
Shape/Size: Shield, 2"x1.75"	Shape/Size: Shield, 2.65"x2.25"
Const/Rarity: Die-cut / C	Const/Rarity: Die-cut / B
Clasp style: Pocket clip	Clasp style: Soldered safety pin
Item Notes: Hand engraved; "Duty" MAA; possibly locally made	Item Notes: Hand stamped, locally made
Author's collection	Author's collection

Image 115	Image 116
Shape/Size: E-T shield, 2.1"x1.25"	Shape/Size: 6-Pt, hollowed ball-tipped star, 2.45"x2.20"
Const/Rarity: Die-cut / C	Const/Rarity: Die-cut / B
Clasp style: Wire pin and "C" catch	Clasp style: Wire pin and fork catch
Item Notes: Marked Sterling; vintage usage	Item Notes: Heavily vaulted (stamped arch), hollowed ball-tips are part of the original die-stamp design. Many ball-tips are the applied type after initial stamp.
Author's collection	Author's collection

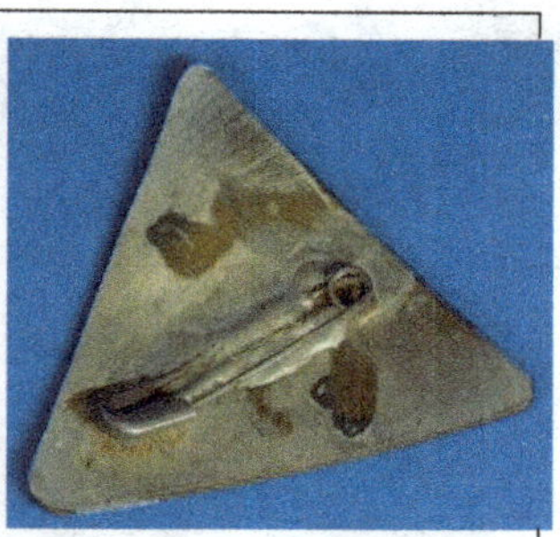

Image 117	Image 118
Shape / Size: 6-Point Star / 2.5"x2.5"	Shape: Triangle shield
Const/Rarity: Die-cut / C	Const/Rarity: Die-cut / C
Clasp style: Spring pin	Clasp style: Soldered safety pin
Item Notes: Early usage	Item Notes: Scarce design

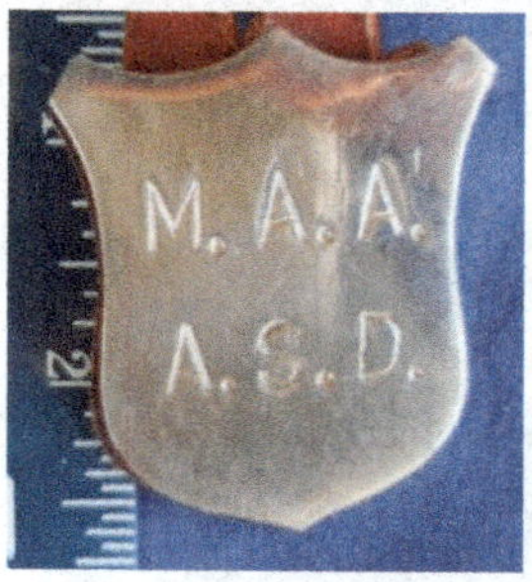

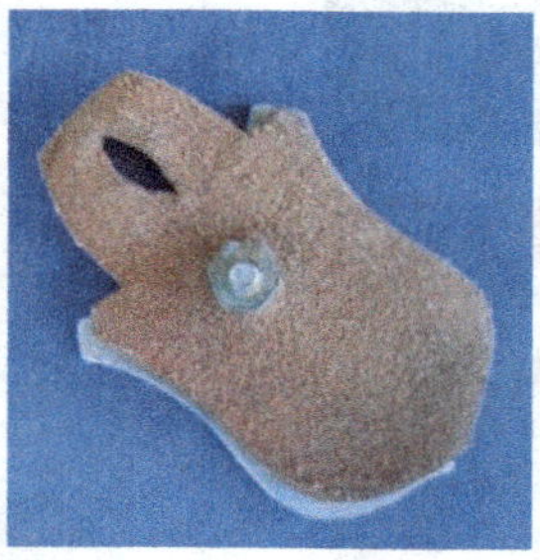

Image 119	Image 120
Shape: Shield	Shape: Eagle top shield
Const/Rarity: Die-cut / A	Const/Rarity: Die-cut / B
Clasp style: Screw back and pocket hanger	Clasp style: Wire pin & small Burgess catch
Item Notes: Possibly Avia. Supply Dept., or Advanced Supply Depot	Item Notes: Hard enamel crest, hallmarked
Images:	Images: David Fox collection

Additional badge examples (Image 121)

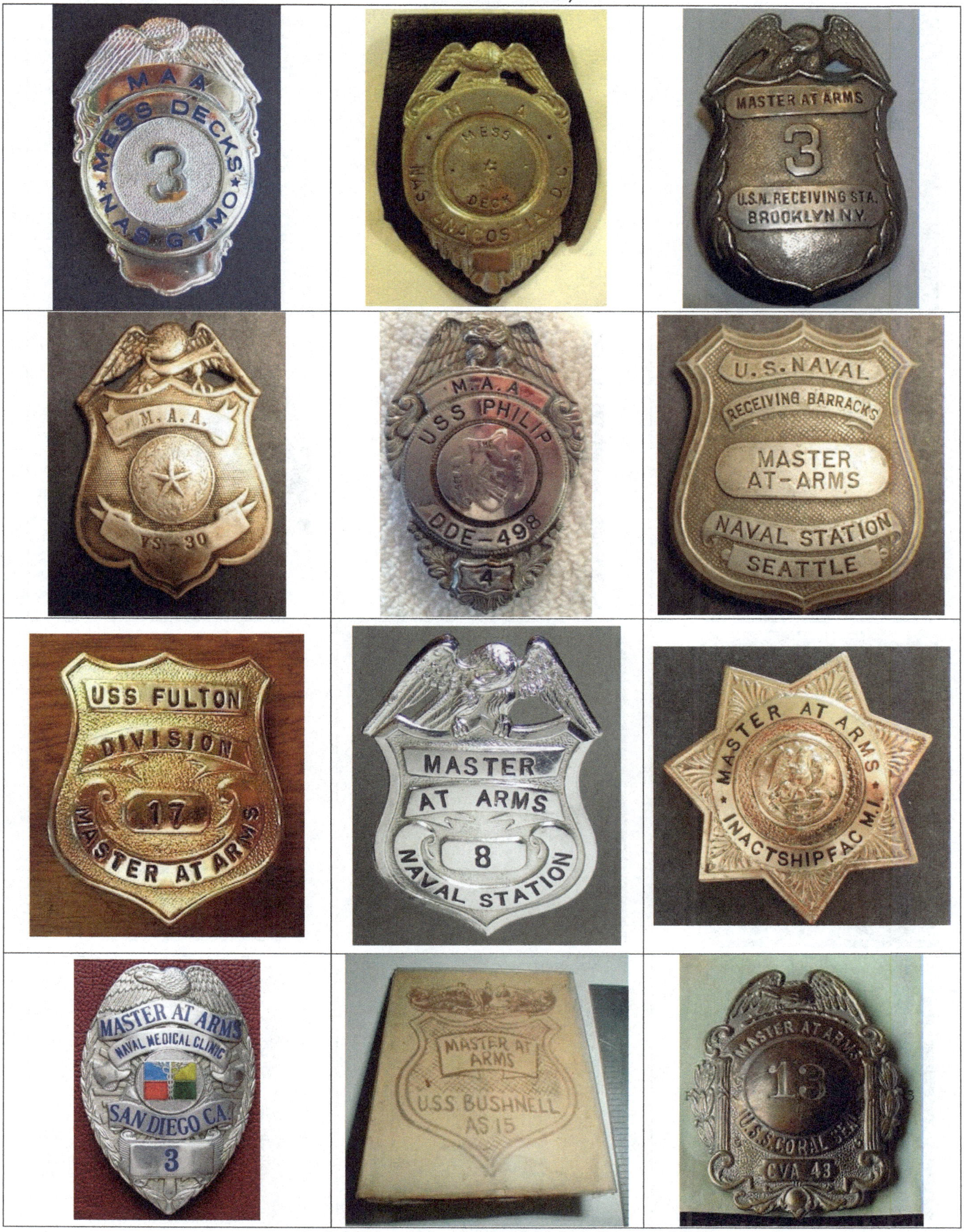

Additional badge examples (Image 122)

Additional badge examples (Image 123)

Police Petty Officer

Police Petty Officers and shipboard-designated Shore Patrol Petty Officers were an implied extension of the command's Master-at-Arms force.

In the United States Navy's [The] Blue Jackets' Manual, the enlisted Petty Officer has been tasked with additional duties that were not expected of non-rated personnel.
The Master-at-Arms (MAA) and the Police Petty Officers are men assigned to maintain order on a ship or station. They are the assistants of the executive officer. Large ships will have a Chief Master-at-Arms (CMAA) with several assistants. Men are assigned to the MAA force for several months or longer, and, while acting as MAAs, are relieved of most of their normal watches and duties. Police Petty Officers are Petty Officers who remain primarily with their divisions for work and watches but are assigned additional duties within their divisions such as maintaining order, enforcing reveille and taps, directing traffic [flow], and turning lights on and off.[viii]

Interviews with Navy veterans who were Petty Officers and who served on active duty between the years 1948 through 1968 have found that the majority of those personnel, especially those who experienced sea duty, did fulfill the duties listed above. However, only one interviewee remembered ever having seen a Police Petty Officer (PPO) badge.

For the most part, PPO badges appear to have been issued only at the larger commands and larger shore stations, especially training centers, naval air stations, and aircraft carriers. An example from the USS Bennington (CVS-20), dated from the early 1960s, is included below.

Image 124
Shape/Size: Eagle-top shield
Const/Rarity: Die-cut / D
Clasp style: Bar pin
Item Notes: This is a confusing generic Police Petty Officer and Master-at-Arms (M.A.A.) denoted badge.
Images:

Police Petty Officer Afloat

Image 125	Image 126
Shape: Eagle-top shield	Shape: Round
Const/Rarity: Die-cut / B	Const/Rarity: Die-cut / C
Clasp style: Wire pin and fork catch	Clasp style: Soldered safety pin
Item Notes:	Item Notes: Early usage, USS Augusta (CA-31)
Images:	Images:

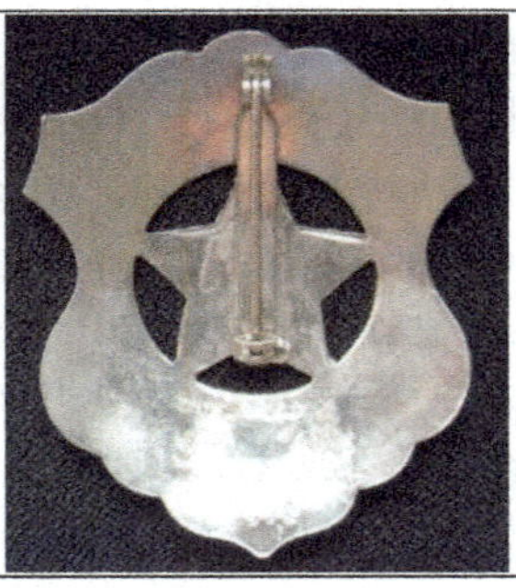

Image 127	Image 128
Shape: Shield w/cut out star	Shape/Size: Eagle-top shield, 2.4"x1.65"
Const/Rarity: Die-cut / B	Const/Rarity: Die-cut / B (in general, PPO badges are not common)
Clasp style: Bar pin	Clasp style: Bar Pin
Item Notes: "M" machinery division PPO	Item Notes: The Grand Canyon was a "Tender"
Images:	Images: Author's collection

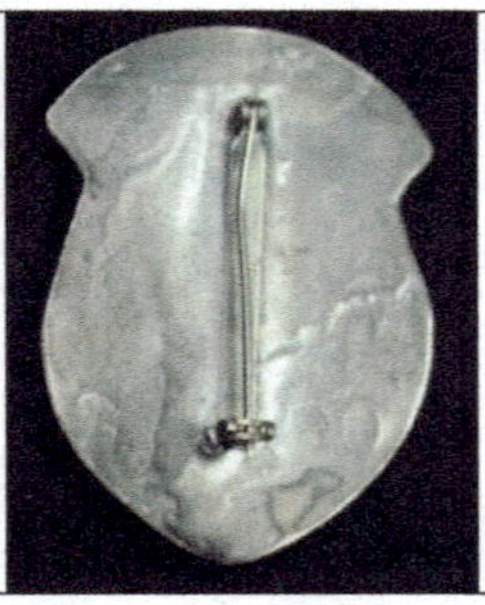

Image 129	Image 130
Shape: Shield	Shape/Size: Pinched shield, 3"x2"
Const/Rarity: Die-cut / A	Const/Rarity: Die-cut / B
Clasp style: Spring pin w/small saddle	Clasp style: Bar pin
Item Notes: "S" division in USN units stands for "Supply" Division. In larger units the "S" is sub-divided by job description, storekeeper, laundry, data processing, disbursing, etc.	Item Notes: Locally made, hand lettered
Images:	Images: Author's collection

Image 131	Image 132
Shape/Size: Flat Sided Shield, 2.6"x1.55"	Shape: Round w/cut-out star
Const/Rarity: Die-cut / B	Const/Rarity: Die-cut / B
Clasp style: Wire pin and small Burgess catch	Clasp style: Bar pin
Item Notes:	Item Notes: "A" division (aux. machinery)
Images: Author's collection	Images:

Image 133	Image 134
Shape: Pinched Shield w/cut-out Star	Shape/Size: Shield, 2.5"x1.7"
Const/Rarity: Die-cut / C	Const/Rarity: Die-cut / B
Clasp style: Wire pin and fork catch	Clasp style: Pin and small Burgess catch
Item Notes: "H" Hospital/Health Division	Item Notes:
Images: David Fox collection	Images: Author's collection

Image 135	Image 136
Shape/Size: Eagle-top shield, 2.4"x1.7"	Shape/Size: Shield w/star, 2.6"x2.15"
Const/Rarity: Die-cut / C	Const/Rarity: Die-cut / B
Clasp style: Spring pin w/saddle	Clasp style: Pocket clip
Item Notes: Early usage, Escort Carrier, only in use from 1943-46.	Item Notes: Pocket clip might not have been original clasp
Author's collection	Author's collection

Image 137	Image 138
Shape/Size: 6-Pt Star, 2.75"x2.4""	Shape/Size: Eagle-top shield, 3"x1.85"
Const/Rarity: Die-cut / B	Const/Rarity: Die-cut / B
Clasp style: Pin and tongue catch	Clasp style: Pin and fork catch
Item Notes: No hallmark	Item Notes: Marked "Sterling"
Author's collection	Author's collection

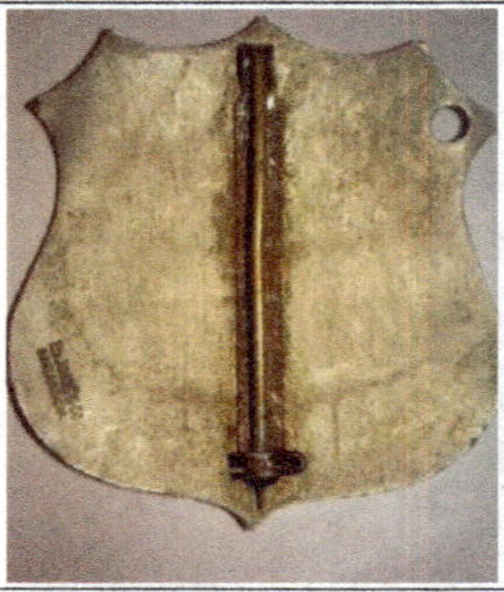

Image 139	Image 140
Shape/Size: Shield, 2.25"x2.05"	Shape: Eagle-top shield
Const/Rarity: Die-cut / B	Const/Rarity: Die-cut / C
Clasp style: Pin and small Burgess catch	Clasp style: Pin and Burgess catch
Item Notes: HM Ed Jones Co, Oakland CA	Item Notes: Multi-layer badge
Author's collection	Images:

Image 141	Image 142
Shape: 6-Pt, Ball-tipped star	Shape: Eagle-top shield
Const/Rarity: Die-cut / C	Const/Rarity: Die-cut / C
Clasp style: Spring pin w/saddle	Clasp style: Pin and tongue catch
Item Notes: Training school PPO	Item Notes: The Yosemite was one of the author's ships; we called her the "YoYo"
Images: David Fox collection	Images: David Fox collection

Police Petty Officer Ashore

Police Petty Officers often worked with the MAA force. This photo is of the Fleet Hospital on Guam in 1944-45. The caption is “MAA Shack,” but six of the personnel are wearing versions of this PPO badge.

MAA SHACK

H. G. Johnson
W. C. Moore
V. L. Fisher
Wallace
Chappelle
Schmitz
Fanger
J. M. Brown
Doran

MAA Shack, Guam, 1944-45, (Image 143), David Fox collection

 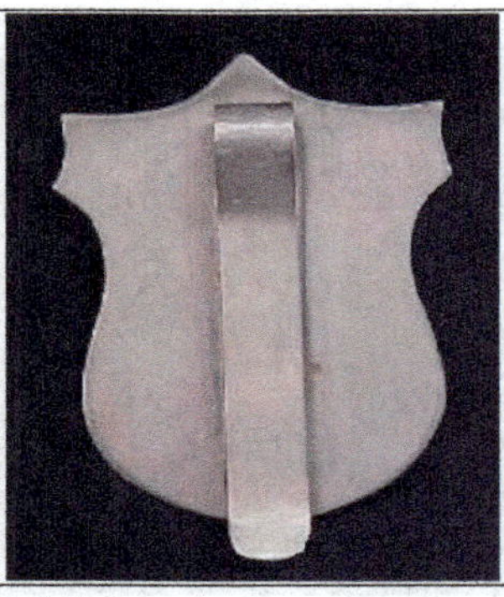	 
Image 144	Image 145
Shape: Pinched shield	Shape/Size: Shield / 2.55”x1.65”
Const/Rarity: Die-cut / B	Const/Rarity: Die-cut / C
Clasp style: Pocket clip	Clasp style: Bar pin
Item Notes: Simple flat shield	Item Notes: This unit name 1943-46 only
Images:	Author’s collection

Image 146	Image 147
Shape: Pinched Shield	Shape: Eagle-top shield
Const/Rarity: Die-cut / B	Const/Rarity: Die-cut / B
Clasp style: Wire pin and fork catch	Clasp style: Spring pin w/saddle
Item Notes:	Item Notes: HM Ed Jones, twice
Images: USMilitariaForum.com	Images:

Image 148	Image 149
Shape: Eagle-top shield	Shape/Size: Pinched Shield, 2.2"x2.05"
Const/Rarity: Die-cut / C	Const/Rarity: Die-cut / B
Clasp style: Pin and small Burgess catch	Clasp style: Spring pin w/saddle (broken)
Item Notes: Asst. Chief PPO, scarce wording	Item Notes: NOB, Naval Operating Base, WWII, HM: Meyer and Wenthe, Chicago
Image: FlyingTigerAntiques.com	Author's collection

Image 150	Image 151
Shape/Size: Eagle-top shield , 2.45"x1.8"	Shape: 7-Point Star
Const/Rarity: Die-cut / B	Const/Rarity: Die-cut / B
Clasp style: Spring pin w/saddle	Clasp style: Bar pin
Item Notes: Great Lakes, Navy "Boot Camp"	Item Notes: Heavily vaulted (arched) badge
Author's collection	Images:

Image 152	Image 153
Shape/Size:6-Pt ball-tipped star 2.45"x2.15"	Shape/Size: Flat Shield, 2.0"x1.6"
Const/Rarity: Die-cut / C	Const/Rarity: Die-cut / D
Clasp style: Spring pin w/saddle	Clasp style: Soldered safety pin
Item Notes: Naval Ordnance Test Station, China Lake, Calif., established in 1943	Item Notes: Oran, Algeria base, WWII only
Author's collection	Author's collection

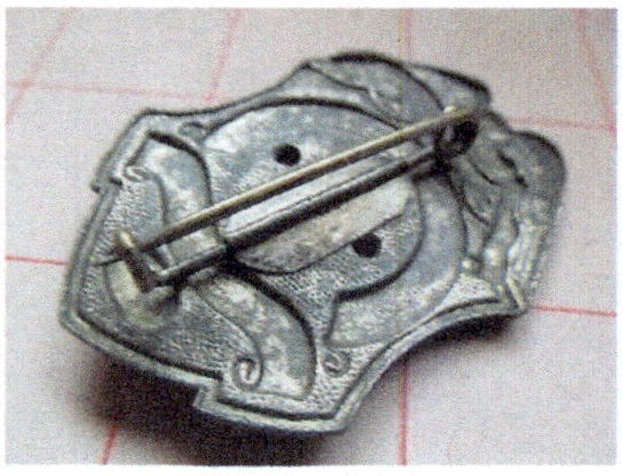

Image 154	Image 155
Shape/Size: Pinched Shield w/cut-out star, 2.1"x1.75"	Shape: Eagle-top pinched shield
Const/Rarity: Die-cut / B	Const/Rarity: Die-cut / C
Clasp style: Spring pin w/saddle	Clasp style: Spring pin w/saddle
Item Notes: Puget Sound Navy Yard, large Washington State naval shipyard	Item Notes: Service school training center; barracks PPO; scarce
Author's collection	Images: Bill Hedges collection

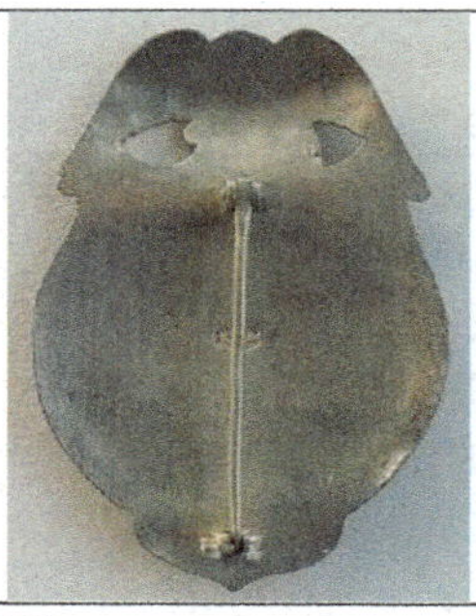

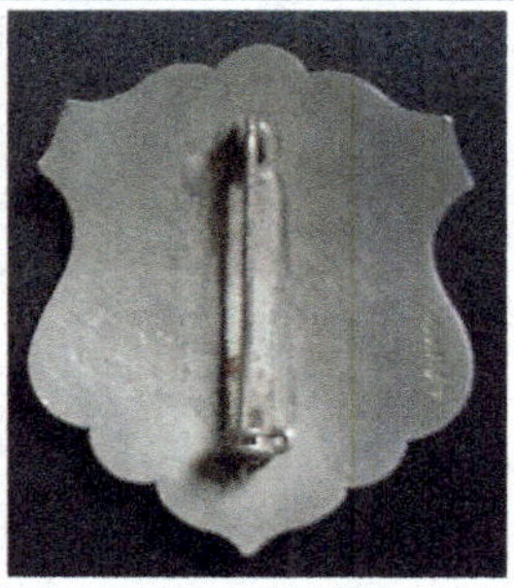

Image 156	Image 157
Shape: Eagle-top Shield	Shape: Pinched shield
Const/Rarity: Die-cut / C	Const/Rarity: Die-cut / B
Clasp style: Pin and small Burgess catch	Clasp style: Bar pin
Item Notes: Barber's Point Terr. of Hawaii	Item Notes: Navy Yard Mare Island. HM.
Images:	Images:

ACORN

The Navy's code word, "ACORN," not an acronym, was used to describe the organization of men and equipment designed to rapidly build and operate new airfield facilities or to rehabilitate captured airfields. The organization always included one or more navy construction battalion units.

A Patrol Aircraft Service Unit (PATSU) was assigned to an ACORN to provide support for attached patrol aircraft units. Where several aircraft types were operating from the same base, an augmented unit known as a Combat Aircraft Service Unit would be provided to the ACORN. Other types of aircraft service units existed and occasionally provided services to patrol aircraft. These land-based facilities became increasingly vital as the build-up of land-based PBY-5A Catalina, PB4Y Liberator/Privateer, and PV Ventura/Harpoon aircraft accelerated. Operating from bases on Wallis, Canton, Nukefetau and Funafuti islands in the Phoenix and Ellis Island groups, missions were flown against Tarawa and Makin in the Gilbert Islands and against Maloelap, Mili Atoll, Jaluit, Kwajalein, Wotje and targets of opportunity in the Marshall Islands.[ix]

Image 158
Shape: Eagle-top shield
Const/Rarity: Die-cut / D
Clasp style: Wire pin and Burgess catch
Item Notes: Rare civilian contractor unit, from WWII, most often used with construction of naval air bases and working with Navy Construction Battalions (CBs) or Seabees; this badge stamped "Sterling" with hard enamel lettering. There is also at least one interesting version inscribed with "Air Force" "Police Petty Officer" and "ACORN."
Image: Unknown

Police Petty Officer (Generic / Unit Unknown)

	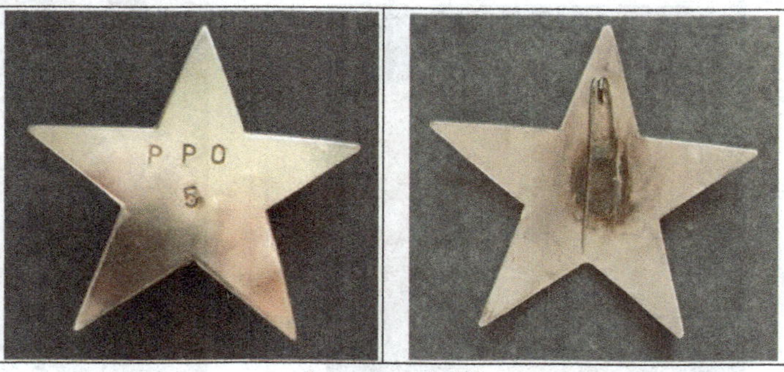
Image 159	Image 160
Shape/Size: 6-Pt ball-tipped star, 2"x1.8"	Shape/Size: 5-Point star, 2.3"x2.4"
Const/Rarity: Die-cut / C	Const/Rarity: Metal snips / C
Clasp style: Pin and small Burgess catch	Clasp style: Usually a safety pin was soldered on these "shop made" badges, looks like the head of the safety pin broke off
Item Notes: Tips are applied, nice presentation	Item Notes: Locally made and hand lettered
Author's collection	Author's collection

Image 161	Image 162
Shape: 6-Point ball-tipped star	Shape/Size: 6-Point star, 3"x2.6"
Const/Rarity: Die-cut / C	Const/Rarity: Die-cut / B
Clasp style: Pin and Burgess catch	Clasp style: Spring pin
Item Notes: Marked Sterling, slight vault	Item Notes: Slight vault
Images:	Author's collection

Image 163	Image 164
Shape: Shield	Shape/Size: Pinched shield, 2.109"x1.75"
Const/Rarity: Plastic-Celluloid / C	Const/Rarity: Die-cut / B
Clasp style: Bar pin w/saddle	Clasp style: Bar pin
Item Notes: Rare plastic, generic PPO badge	Item Notes: Slight vault, Author's Collection

Additional badge examples (Image 165)

All Images Bill Hedges Collection (Image 166)

Shore Patrol

While not stated as such in U.S. Navy regulations, Police Petty Officers and shipboard designated Shore Patrol Petty Officers were an implied extension of the command's Master-at-Arms force as far as shipboard personnel were concerned.

Unit-supplied shore patrol duty was assigned to Petty Officers over the age of 21, usually when a ship entered a port other than that ship's homeport. This did not happen in the larger U.S. ports where a Navy presence was routine. For example, the ships on which the author was stationed did not assign shore patrol (SP) duty when the ship pulled into San Diego, California, or Norfolk, Virginia. However, shore patrol duty was assigned in, for example, Rio De Janeiro, Brazil, where the author was assigned to a car with two local police officers, and in Valetta, Malta, in the Mediterranean Sea, where SP duty was assigned alongside two British sailors. At that time, the British had a permanent naval base on the island of Malta.

In most naval districts, a standing shore patrol or permanent shore patrol was in place. In the case of foreign ports, the permanent shore patrol was often augmented with the temporary duty shore patrol members from visiting naval units.

Established in 1942, the Specialist (S) naval rating (Shore Patrol and Security), filled shore patrol teams and ensured basic shore station security or ship-supplied shore patrol members. The Specialist (S) name was changed in 1948 to Shore Patrolman, and it took on some of the official functions of the current Master-at-Arms rating, only to have Specialist (S) rating be once again disestablished in 1953. At that time, Navy shore patrol duties were once again assigned to other ratings as temporary duties, even if the duty was in a naval district with permanent shore patrol assignment. In this case, the rating specialty was something other than Specialist (S); it was often a Boatswains Mate, Gunner's Mate or Aviation Machinist's Mate, for example.

Usage of badges other than arm brassards by USN shore patrol personnel was forbidden in the 1943 Shore Patrol manual.[x]

In July 1945 a shore patrol metal badge was authorized and limited to permanent shore patrol members who were grouped by Naval Districts. The naval district number was often included in the shore patrol badge design. With this same uniform change, the SP belt and leggings colors were changed from khaki to white.

In the 1953 edition of the U.S. Navy Shore Patrol manual, there was no longer a reference to a metal badge for permanent shore patrol personnel.[xi]

Because Shore Patrol badges were only issued for a short number of years to permanent shore patrol members and were highly controlled equipment, SP badges are the most difficult for a collector to obtain. Metal Shore Patrol badges should all be considered scarce or rare.

Since their origin in 1913, Naval districts have been established, and many have subsequently been disestablished over the years.

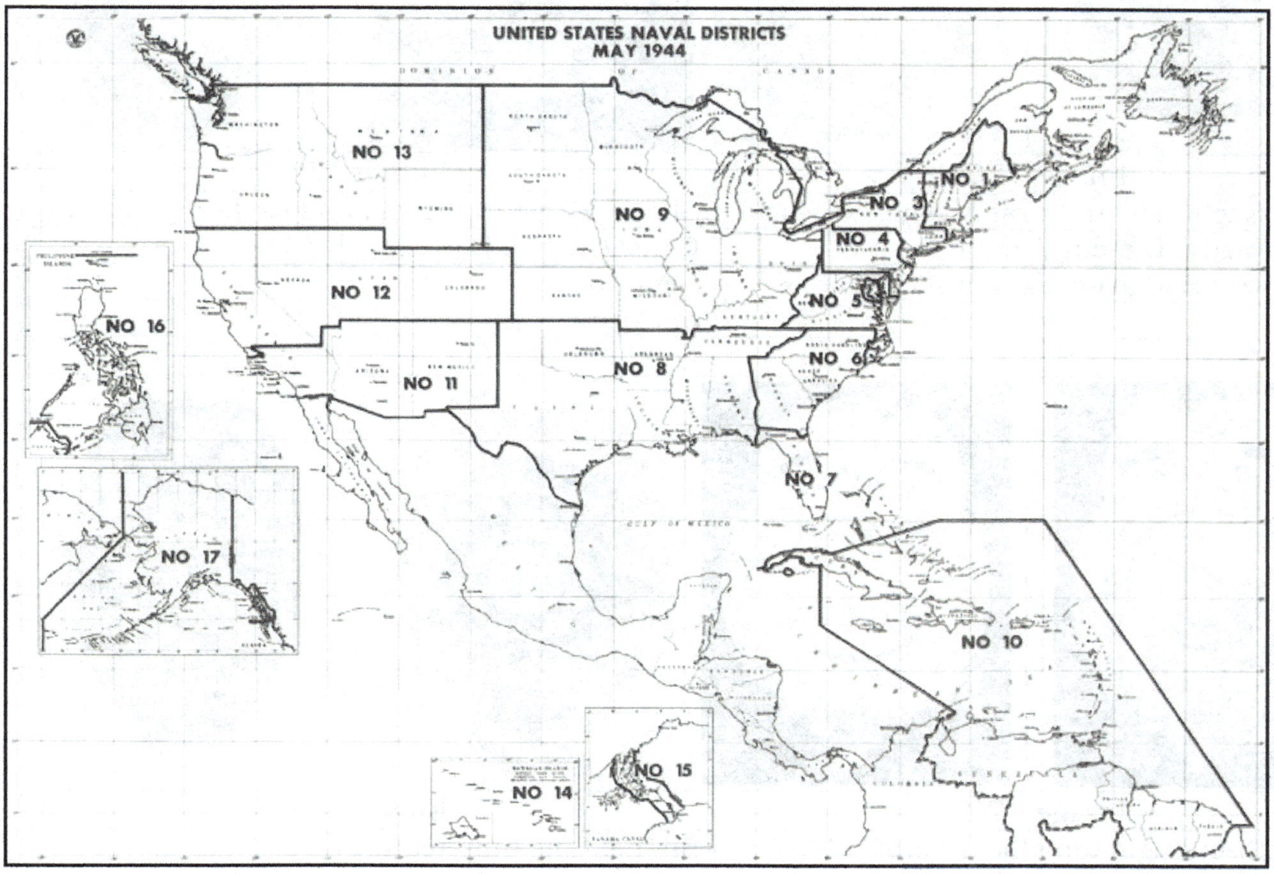

(Image 167)

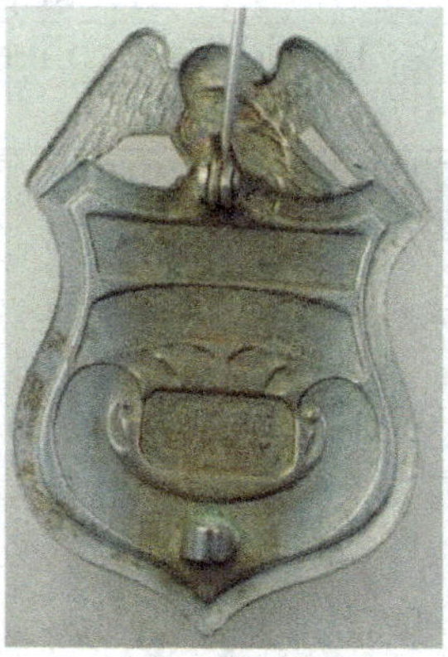

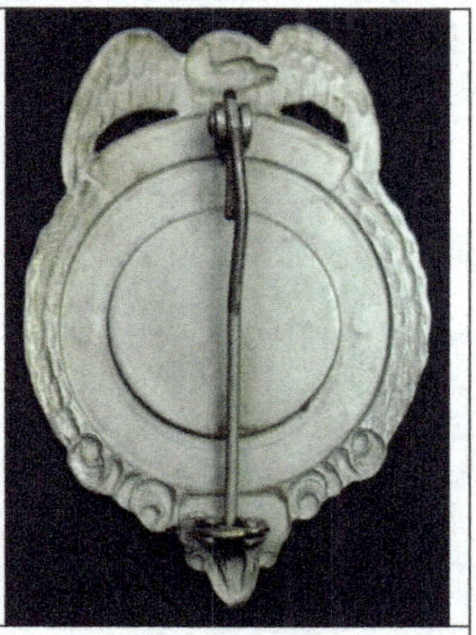

Image 168	Image 169
Shape/Size: Eagle-top shield, 2.45"x1.7"	Shape/Size: Eagle-top shield, 2.55"x1.75"
Const/Rarity: Die-cut / D	Const/Rarity: Die-cut / D
Clasp style: Pin and tongue catch	Clasp style: Pin and fork catch
Item Note: HM F.G. Clover Co. New York, NY	Item Notes:
Author's collection	Author's collection

Image 170	Image 171
Shape: Eagle-top sunburst & pinched shield	Shape: Eagle-top shield and eagle-top shield
Const/Rarity: Die-cut / E / E	Const/Rarity: Die-cut / E / D
Item Notes: Badge on right is very rare	Item Notes:
Images:	Images:

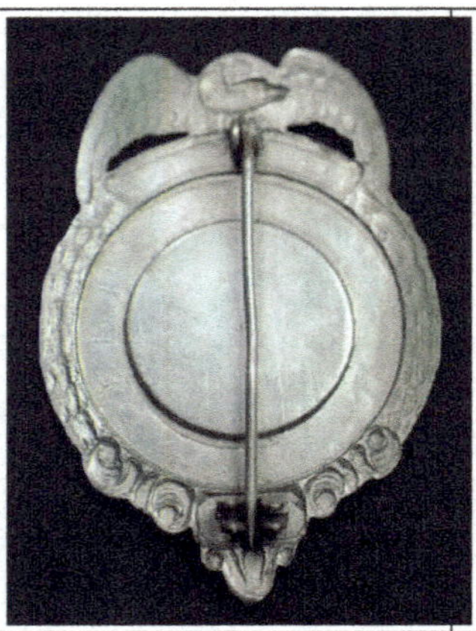

Image 172	Image 173
Shape/Size: Eagle-top shield, 2.55"x1.75"	Shape: Eagle-top teardrop
Const/Rarity: Die-cut / D	Const/Rarity: Die-cut / D
Clasp style: Pin and fork catch	Clasp style: Pin and small Burgess catch
Item Notes:	Item Notes: S.P. Investigator, hallmarked
Author's collection	Images: David Fox collection

Image 174	Image 175
Note: Florida State seal is interesting	Note: Rare 5th Naval District badge
Images:	Images:

(Image 176)

(Image 177)

Petty Officer Barry L. Lease, shown above left, Naples, Italy, 1963-66. The badge Petty Officer Lease is wearing is not a shore patrol badge but a NATO Allied Forces Southern Europe badge worn by all military personnel stationed in Southern European countries at that time. Photo on the right is of S.P. in 1945 (note S.P. badge and armband).

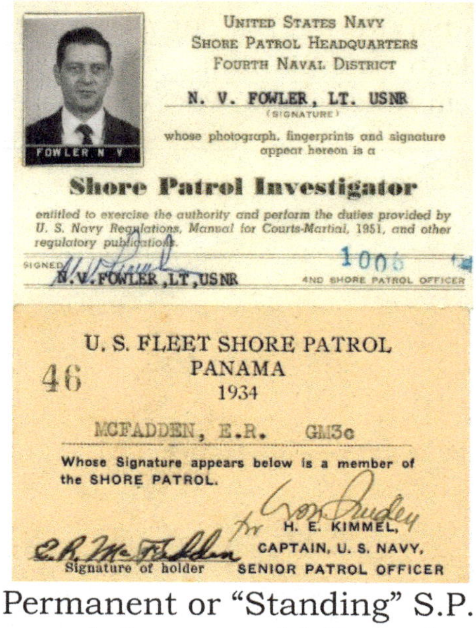

FOWLER N V

UNITED STATES NAVY
SHORE PATROL HEADQUARTERS
FOURTH NAVAL DISTRICT

N. V. FOWLER, LT. USNR
(SIGNATURE)

whose photograph, fingerprints and signature appear hereon is a

Shore Patrol Investigator

entitled to exercise the authority and perform the duties provided by U. S. Navy Regulations, Manual for Courts-Martial, 1951, and other regulatory publications.

SIGNED N.V. FOWLER, LT, USNR 1005 4ND SHORE PATROL OFFICER

U. S. FLEET SHORE PATROL
PANAMA
1934

46

MCFADDEN, E.R. GM3c

Whose Signature appears below is a member of the SHORE PATROL.

Signature of holder

H. E. KIMMEL,
CAPTAIN, U. S. NAVY,
SENIOR PATROL OFFICER

Permanent or "Standing" S.P. carried a second I.D. card. David Fox collection (Image 178)

David Fox collection (Image 179)

Navy Use of Guards, Civilian Contract Police, and Security Forces

Masters-at-Arms are Navy personnel. Most U.S. Navy police are civilian contractors working on a naval installation. At one point in the early 2000s, the Navy used Master-at-Arms personnel, badged as "Police" on some bases.

U.S. Navy personnel working as badged guards are mostly from the WWII era. These personnel worked warehouse and shipyard security, fire watches and brig guards.

U.S. Navy personnel badged as "Security" forces have a more recent history working missile sites and other areas deemed as high targets. They can be either civilian contractors or active USN personnel.

The job descriptions of guards and security personnel were often the same, but whether the badge was worn by civilian contractor or U.S. Navy personnel was mostly a matter of the timing of often changing USN directives.

Image 180	Image 181
Shape/Size: Eagle-top shield, 2.55"x1.8"	Shape: Eagle-top shield
Const/Rarity: Die-cut / B	Const/Rarity: Die-cut / B
Clasp style: Pin and fork catch	Clasp style: Bar pin
Item Notes: Dock guard attached to the Industrial Manager, 8th Naval Dist., New Orleans 1942-1945	Item Notes: Naval Post Graduate School
Author's collection	Images:

Sampson Naval Training Station contract police, WWII (Image 182)

 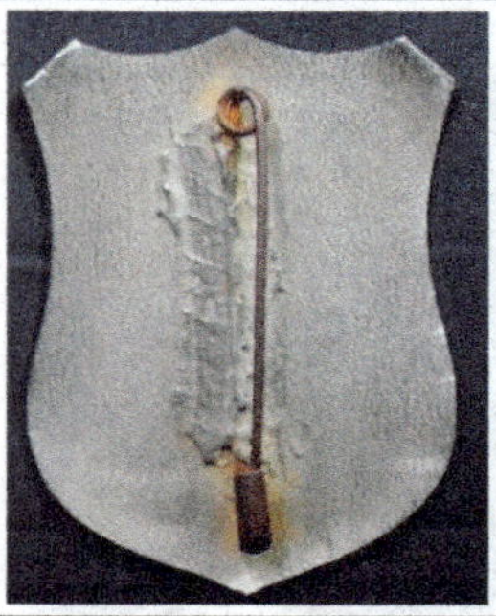	 
Image 183	Image 184
Shape/Size: Pinched shield, 2.5”x1.75”	Shape/Size: Eagle-top Shield, 2.5”x1.7”
Const/Rarity: Die-cut / C	Const/Rarity: Die-cut / B
Clasp style: Soldered safety pin	Clasp style: Pin and large Burgess catch
Item Notes: Locally made, hand stamped, brig “chaser” wording on badge usage is scarce; the job consisted of escorting personnel from civilian jail or one brig to another.	Item Notes: Brig staff member, Charleston, S.C.
Author’s collection	Author’s collection

Image 185	Image 186
Shape: Eagle-top shield	Shape: Eagle-top shield
Const/Rarity: Die-cut / A	Const/Rarity: Die-cut / A
Clasp style: Bar pin	Clasp style: Bar pin
Images:	Image: FlyingTigerAntiques.com

Image 187	Image 188
Shape: Eagle-top shield	Shape: Eagle-top shield
Const/Rarity: Die-cut / B	Const/Rarity: Die-cut / B
Clasp style: Bar pin	Clasp style: Pin with tongue catch (broken)
Images: David Fox collection	Images: David Fox collection

Image 189	Image 190
Shape/Size: 5-Pt ball-tipped star, 2.65"x2.7"	Shape: Eagle-top shield
Const/Rarity: Die-cut / C	Const/Rarity: Die-cut / A
Clasp style: Wire pin and fork catch	Clasp style: Pin and safety catch
Item Notes: Contract Pacific Naval Air Base police, WWII	Item Notes: 1960s
Author's collection	Images:

Image 191	Image 192
Shape/Size: 5-Pt, ball-tipped star, 2.65"x2.65"	Shape: Pinched shield
Const/Rarity: Die-cut / C	Const/Rarity: Die-cut / E
Clasp style: Pin and Burgess catch	Clasp style: Soldered safety pin (repair)
Notes: 1970s Great Lakes, IL area. Detective rank, scarce.	Notes: Earliest USN usage, very rare, 1920s-1930s
Author's collection	Image: Bill Hedges collection

Image 193	Image 194
Shape/Size: Eagle-top, X-shield, 2.35"x2"	Shape: Eagle-top sunburst
Const/Rarity: Die-cut / A	Const/Rarity: Die-Cut / B
Clasp style: Screw post and alignment pin	Clasp style: Pin and small Burgess catch
Notes: Hat usage	Notes: Interesting star crest
Author's collection	Image:

Image 195	Image 196
Shape: E-T shield / E-T shield	Shape: E-T shield / E-T shield
Const/Rarity: Die-cut / C / B	Const/Rarity: Die-cut / A / B
Images:	Images:

Image 197	Image 198
Shape: Pinched shield	Shape/Size: 6-point, ball-tipped star, 2.15"x1.9"
Const/Rarity: Die-cut / C	Const/Rarity: Die-cut / B
Clasp style: Spring pin	Clasp style: Spring pin w/saddle
Notes: Slight vault, early usage	Notes: HM-Tanner Co.
Image: FlyingTigerAntiques.com	Author's collection

Image 199	Image 200
Shape: Eagle-top shield	Shape: Shield w/soldered letters
Const/Rarity: Die-cut / A	Const/Rarity: Die-cut / B
Clasp style: Bar pin	Clasp style: Screw back
Notes: New York Naval Shipyard Police	Notes: Hat badge, Navy Yard Police
Images:	Images:

Image 201	Image 202
Shape: Shield	Shape: Pinched shield
Const/Rarity: Die-cut brass / C	Const/Rarity: Die-cut / C
Clasp style: Soldered safety pin	Clasp style: Pin and fork catch
Notes: Oahu, Hawaii Naval Air Station	Notes: U.S. Naval Academy, Yard Police
Images: FlyingTigerAntiques.com	Images: FlyingTigerAntiques.com

Image 203	Image 204
Shape/Size: Eagle-top shield / 2.5"x1.65"	Shape/Size: Shield / 2.35"x1.65"
Const/Rarity: Die-cut brass / B	Const/Rarity: Die-cut / B
Clasp style: Pin and tongue catch	Clasp style: Wire pin and tongue catch
Notes: HM-Henry Moss & Co.; WWII usage	Notes: NAS Jacksonville, Fla. Insp. & Survey
Author's collection	Author's collection

	S.F.P.D. Special Police badge issued to a Military Police Officer, U. S. Navy, who worked side-by-side with a regular S.F.P.D. officer. The officer's initials appear on the badge. Made by Irvine & Jachens, San Francisco, and dated on the reverse 8-7-1931. Sterling silver. The special police were a separate, neighborhood police watch authorized in the S.F. City Charter.
Image 205	
Shape: 6-point Star	
Const/Rarity: Die-cut / D	
Clasp style: unknown	
Photo: Jim Casey, www.PriceGuide.com	

The Navy sometimes combined their base security forces with both USN Master-at-Arms personnel and civilian contract police officers working together.

This is a unique Police brassard from King's Bay, Georgia, naval base.

David Fox collection (Image 206)

When collecting civilian contract police badges, some collectors prefer to collect badge "sets," consisting of both a breast badge as well as the hat badge.

David Fox collection (Image 207)

U.S. Navy Security forces are a development of the last quarter century. They can be either civilian contractors or active USN personnel. Early security watches included mundane duties such as pier watches, barracks watches, gate guards, and other "fire watch" duties. Current USN security watches are more often guarding and patrolling base perimeters and nuclear components and weapons/missile sites that are perceived as being high threat locations. USN Security badges cover a very wide range of usage applications.

Image 208	Image 209
Shape: Eagle-top shield	Shape: Shield
Const/Rarity: Die-cut / B	Const/Rarity: Die-cut / B
Clasp style: Pin and fork catch	Clasp style: Pin and Burgess catch
Item Notes: Security Patrol, USS Boston CG-1, 1950s-60s	Item Notes:
Images:	Images:

Image 210	Image 211
Shape: Eagle-top shield, eagle-top shield	Shape: Eagle-top teardrop and sunburst
Const/Rarity: Die-cut / A	Const/Rarity: Die-cut / A
Item Notes: The strips holding lettering for "Supervisor," "FLTWING Security," "NAS Oceana" and "VA" are called panels.	Item Notes:
Images:	Images: David Fox collection

(Image 212)

The Navy Nuclear Weapons Security Insignia (NWS) is a Navy qualification badge that is awarded to enlisted sailors assigned to a Strategic Weapons Facility as base highest level security personnel. The insignia represents completion of 12 months of fully-qualified service to the NWS mission. A gold star mounted at the bottom of the insignia indicates an additional year of qualified service to the NWS mission. A silver star indicates five years of fully-qualified service.

Additional badge examples (Image 213)

Crests Found on USN Law Enforcement-Style Badges

While most badge collectors refer to the additions below as seals, U.S. Navy personnel refer to these items as crests. The addition of a separate crest to a badge makes that badge a two-layer badge. Approximately half of the crests below were applied after the initial badge "blank" was die-cut or die stamped.

(Image 214)

Badge Materials and Construction

Early process, most of which is still used currently:

George E. Virgines, author of Badges of Law and Order, describing early badge production wrote,

[A]*t least nine different processes are required to complete a [custom] badge. First, it must be designed...* [S]*econd, an expert die-cutter cuts by hand every line, letter and figure to assure a precise and finely detailed badge. Third, the new die is placed in a huge stamping press and the badge blank is plated in the machine, and under tons of pressure the piece of flat metal becomes a badge. Next comes trimming of excess metal with precision cutting tools. Skilled lettering men must carefully form the letters clear and sharp. Enameling is carefully filled in with a smooth finish, then, it is baked for a rich looking permanent finish.*

The next process is to fix fastenings and seals into permanent position. Polishing the metal is necessary before the badge is plated. Several finishes can be applied, gold plate, nickel plate and the new rhodium plate. Before this can be done the badge is copper plated, followed by a hot-nickel and then final finish.[xii]

Virgines also defined many of the unique terms used in the construction of, in this case, law enforcement-style badges:

[A] *variety of terms was used to designate the finish of a badge...gold,...10 karat and 14 karat are most commonly used. Gold filled, upon a thick sheet of base metal (brass, nickel, oreide or even sterling silver) a thinner sheet of gold is hard soldered. Then the sheet, composed of the two metals, is rolled down to the required thickness.*

Sterling silver is a fine silver to which has been added 7 ½ 0/0 copper alloy to give hardness, pure silver is extremely soft. Sterling is used in the best silver badges and takes hard enamel [paint for lettering] well.

Nickel is widely used on cheaper badges; it stands up well, usually a soft enamel [for filling in lettering] is used with nickel.

Oreide is a bronze with a high copper content. It wears well, and has the appearance of gold if kept polished. It is the best material for gold plate and takes hard enamel well.

Brass, if used, is usually soft enameled and nickel plated. It cannot be hard enameled.

German Silver is a white alloy of copper, with zinc, and nickel. It is long lasting and takes a soft enamel. It was used for many of the early badges.

Hard enamel – the powdered enamel is placed in the depressed portions of the badge. Then the badge is oven-fired, which fuses the enamel to the metal. When cooled, the enamel is stoned off level with the surrounding metal. This leaves the letters filled with a smooth, shiny, glasslike substance, the same as used on the highest class of jewelry.

Soft enamel – the face of the badge is covered with a paint like substance. Before it hardens, the face of the badge is carefully cleaned off, leaving the depressed portions filled with whatever the design calls for.

Stamped inscription – letters are stamped [into the badge blank] during manufacturer. This is the best, deepest and least expensive form of inscription.

Engraved inscription – letters are hand [or machine] engraved into the badge blank. This is [generally] done after the badge is finished... [but before final plating.][xiii]

Early badge hand engraving (Image 215)

War Material Shortages

During World War II, the United States declared that a metal shortage existed due to the metal requirement for many war efforts.

Manufacturers of civilian badges and most military medals either temporarily ceased production or switched to other, less critical, or more plentiful materials.

Badge manufacturers also produced badges using zinc, plastics, celluloid, and other materials other than metal. Below are a few examples.

Image 216

Image 217

Foreign Made

In the U.S. Navy, a number of badges were produced by any method or producer that was available to ships while they were on patrol and cruises or when they were near permanent overseas shore bases.

For example, many MAA badges exist that were produced in the Philippine Islands. For the European Theatre, examples of German-produced MAA badges also exist. It is believed the USS Fox badge below was produced in the Persian Gulf area.

<table>
<tr><td>
</td><td>
</td><td>
</td><td>
</td></tr>
<tr><td colspan="2">Image 218</td><td colspan="2">Image 219</td></tr>
<tr><td colspan="2">Item Notes: Ramsey possibly foreign made, 1967-88 ship, w/late tongue catch. The USS W.H. Standley was re-designated CG-32 in July 1975 and was decommissioned in 1994.</td><td colspan="2">Item Notes: The USS Fox was re-designated CG-33 in July 1975. Schierstein Base was the U.S. Navy's Rhine River Patrol in the 1950s.</td></tr>
</table>

Navy Unit Produced (Shipboard or Shore Station Shop)

Navy repair ships, also called "tenders," were equipped with machine shops and qualified personnel who easily matched many of the civilian machine shops of the time. These shops and personnel had the skills required to produce non-complicated MAA and PPO badges. Repair shops at Navy shipyards also had the machinery and skilled personnel to produce different quality levels of MAA or PPO badges. If provided "blanks," even smaller ships' personnel could hand-stamp badges.

<table>
<tr><td>
</td><td>
</td><td>
</td><td>
</td></tr>
<tr><td colspan="2">Image 220</td><td colspan="2">Image 221</td></tr>
</table>

Current Production Process

Dave Arnold at BadgeCollecting.com, writing in 2010, after George Virgines, described the following steps for current badge production: "[F]*rom 1902 to today, the shield is made of nickel silver. It is first die-struck in a drop hammer which gives it impression and then is pounded until the desired height of the characters is reached. The perimeter is embossed and the numbers and lugs are hard soldered. The badge is then dipped, cleaned, nickel-plated and polished to a high gloss.*"

Image 222	Image 223
Item Notes: Badges produced in the 1980s and 1990s	

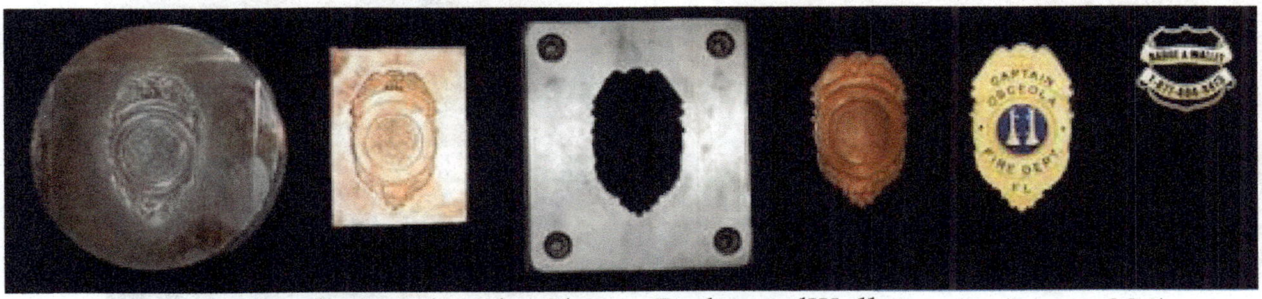

Badge dies and tools, Photo: BadgeandWallet.com (Image 224)

Production Process	
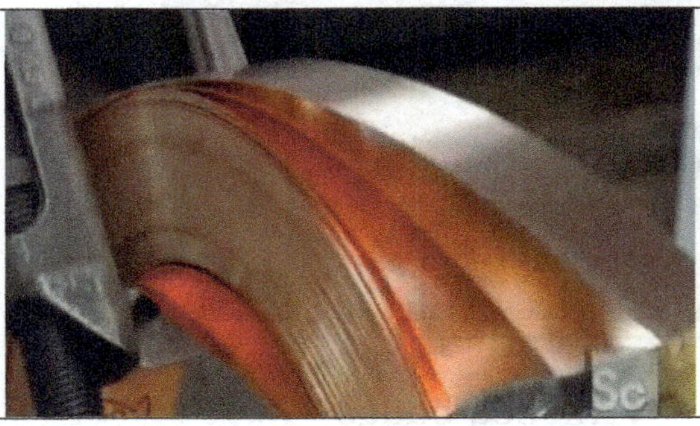	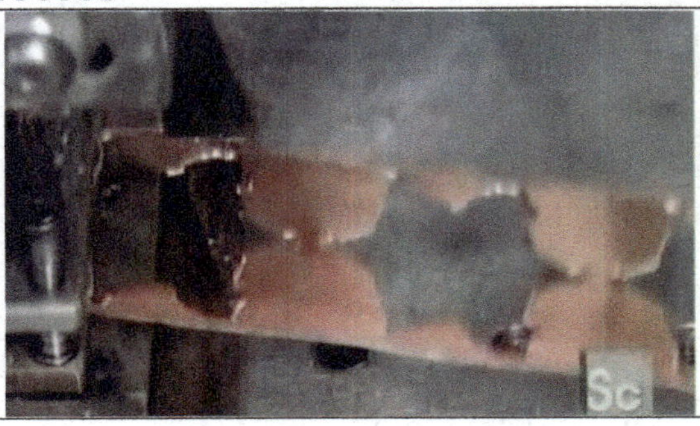

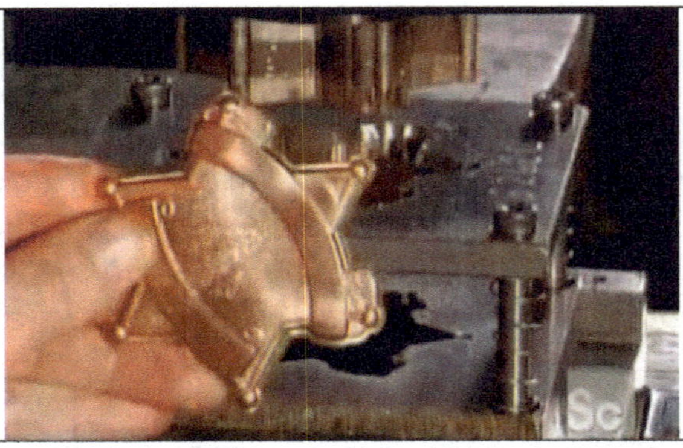	
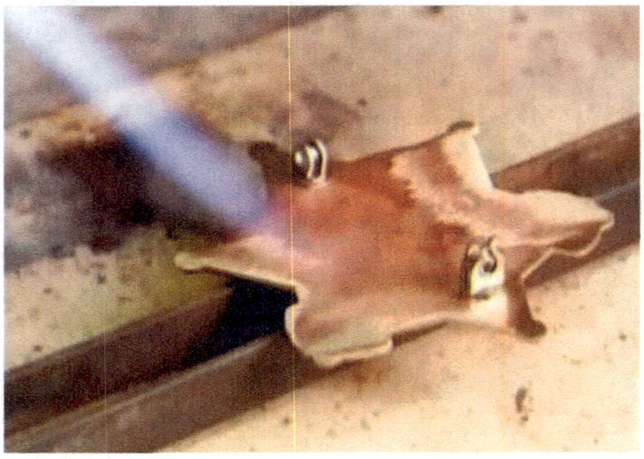	
	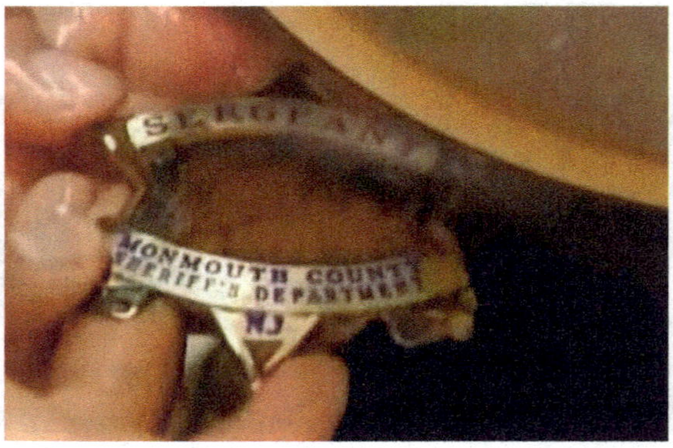

Adding this disk to the original stamped blank makes this a 2-layer badge	Finished badge

(Image 225) Smith-Warren Company (The Science Channel)

Methods of Wear

Suspended

(Image 226)

(Image 227)

(Image 228)

Clipped

(Image 229)

(Image 230)

(Image 231)

Pinned

(Image 232)

(Image 233)

(Image 234)

Wallet Badge

Rare USN wallet-displayed police badge (Image 235)

USS Wright MAA Division, 1967 (Image 236)

Security Uniforms, 1990 (Image 237)

Other USN Badge Usage

More than any other branch of service, the United States Navy utilized metal badges to identify more than just law enforcement positions. Most dealt with positions in Navy shipyards and other Navy shore establishments. To name a few, these included titles such as Lead Laborer, Ship's Crew member, civilian worker ID badge, and firefighter. Here are some examples of these diverse badges.

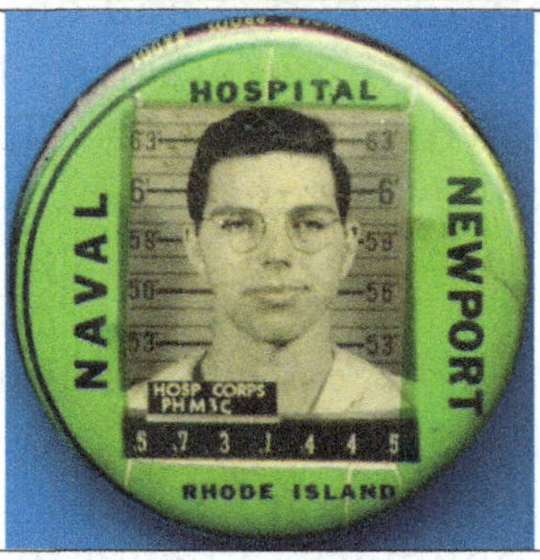

(Image 238)

Hallmarks

The presence of a hallmark on a badge is not necessarily a requirement but might be considered a bonus find for a collector.

A hallmark or maker's mark is usually a name or symbol stamped into the reverse of the badge. A hallmark on a badge provides a number of benefits to the collector. First, depending on the text or symbol of the hallmark, it can help date the badge and obviously tell who the maker was. Second, in some cases, it is possible to determine whether the company had a contract to make badges for the department shown on the badge. This can be key in establishing the authenticity of badges with little or no provenance. (Source: Dale:badgevalues.com)

There have been more than 1,000 unique, documented hallmarks for badges by almost 800 different badge producers. (Source: Dale:badgevalues.com)

Below are detailed photographs of a small sampling of makers' hallmarks known to be used by badge providers to U.S. Navy units.

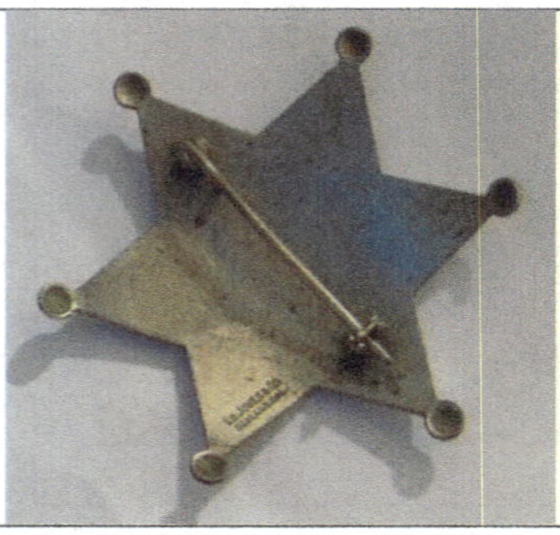	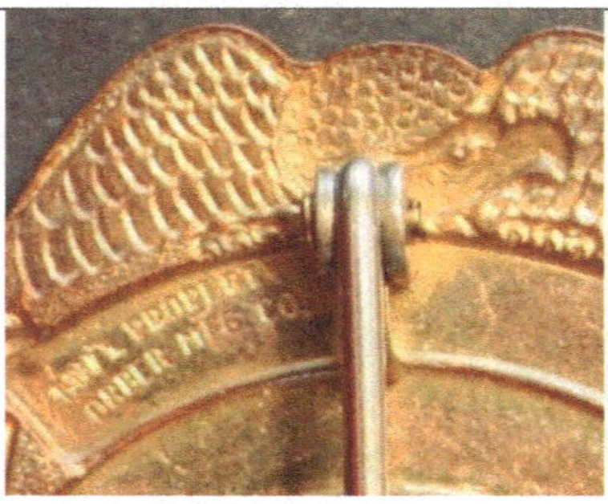	
Image 239	Image 240	Image 241
Hallmark: Ed Jones, Co., Oakland, CA	Hallmark: Orber Mfg. Co.	Hallmark: S.H. Reese, N.Y.

Image 242
HM: F.G. Glover Co., New York, N.Y.

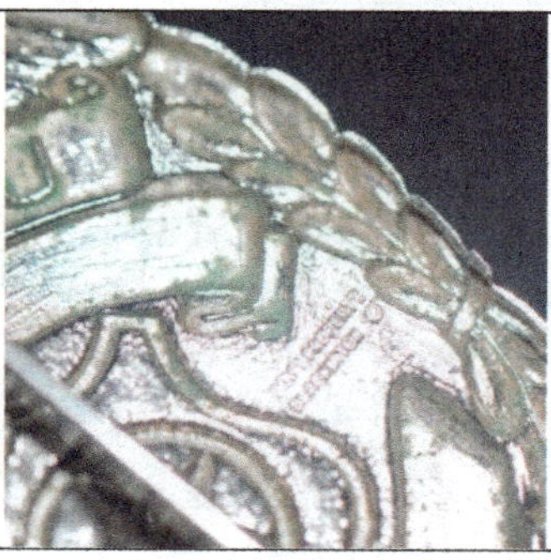

Image 243
Hallmark: Blackinton

Image 244
Hallmark: Cal Stamp Co

Image 245
HM: P.C. Stamp Works

Image 246
HM: N.C. Walter & Sons, 15 Park Row, New York

Image 247
HM: Goldmark Specialty Co, Norfolk

Image 248
HM:N.C. Walter & Sons, NYC (variant)

Image 249
HM: Patrick & M.K. Co., San Francisco

Image 250
HM: Sun Badge Co., L.A. Co.

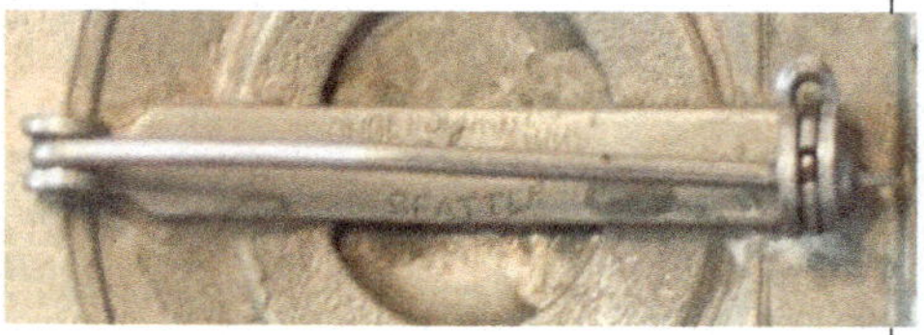

Image 251
HM: Puget Sound

Image 252
Hallmark: Russell Uniform Co.

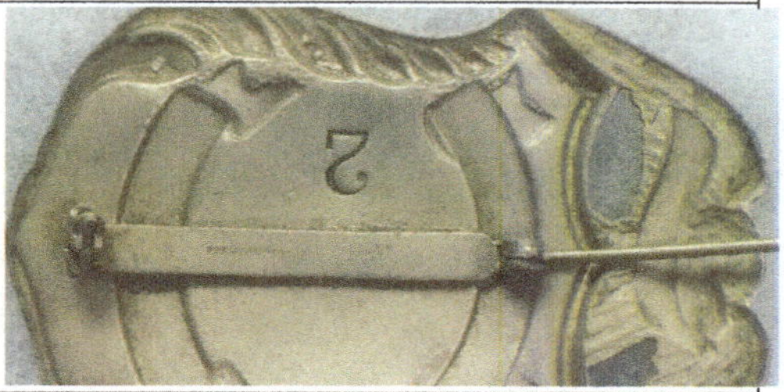

Image 253
Seen on most bar pins, this patent number, 2066969, is for the pin attachment and not the badge, first Pat. in 1937.

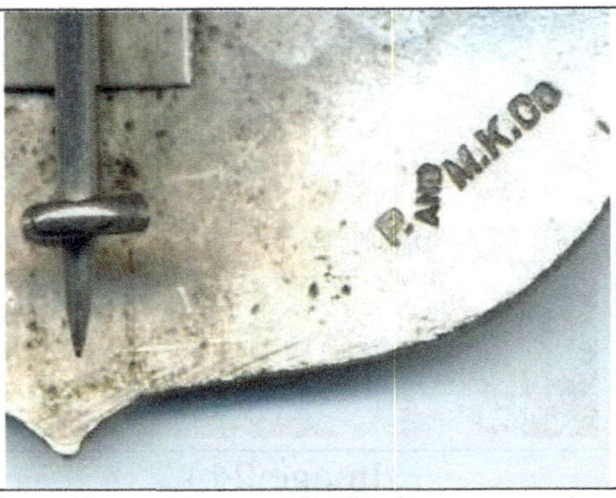		
Image 254	Image 255	Image 256
HM: Patrick & M.K. Co., (variant)	HM: W.S. Darley & Co.	HM: Govt Property, Blackinton

In his book *Law Enforcement Memorabilia*, Monty McCord published **Stephen Wayne Knight's** comprehensive work on badge hallmarks. The following is a sample page.

Hallmark Listing

Maker's Mark (as appears on badge)	Location	Year(s)	Additional	City/State
HEEREN BROS. & CO.	Pittsburgh, PA	Unkn.	Name HM	——
HELWIG & CO. LOUIS	205 W. Madison	1937	——	Chicago, IL
HELWIG & CO. WM.	434 Elm	1937 & 49	——	Cincinnati, OH
HELLENBUSH	Cincinnati, OH	1900-Unkn	Badges	——
HENRY MOSS & CO.	113-123 53rd	1922-1948	——	Brooklyn, NY
HENRY MOSS & CO.	115 53rd St.	1944	——	Brooklyn, NY
HENRY MOSS & CO.	——	1922-1938	Another Branch	Philadelphia, PA
HENDERSON AMES CO.	——	1950s	State Badges	Kalamazoo, MI
HEROLD, GEO.	7 Centre Mkt.	Unkn.	——	New York, NY
HICKOX-HIXSON BADGE	——	Unkn.	Unkn.	El Paso, TX
HI-GLO	——	1978-Present	BLACKINTON HM	Attleboro, MA
HISS COL'S	——	1922	HISS STAMP CO.	Columbus, OH
HISS-TOLEDO	——	1917-1922	Moved	Toledo, OH
H.L. SCHWARTZ & SONS MFG. CO.	——	1939	——	Benton Hbr., MI
H. MORTON & CO.	14th & Broadway	1915-1916	Custom Badges	——
H.M. NUTTER CO.	425 Montgomery St.	1924-1944	'18 Moise salesman	——
H.M. NUTTER CO.	Died '44, Wife operated	1944-1955	S. F. CAL., HM	San Francisco, CA
H.M. NUTTER CO.	S. F. CAL	1944-1955	To PATRICK in '55	San Francisco, CA
H.O. BATES	255 N Broad St.	1939	——	Elizabeth, NJ
HODGES BADGE CO.	166 Essex St.	1937 & 39	——	Boston, MA
HOOK-FAST	220 Eddy	1949	No Mfg. HM	Providence, RI
HOOK FAST SPECIALTS	63 Seymour St.	1992-Present	——	Providence, RI
H.R. TERRYBERRY	Grand Rapids	1939	Michigan	Grand Rapids, MI
H.T. COOK & CO.	——	1852-1887	Died, Co. Sold	Toledo, OH
HUEBEL MFG. CO.	103 Monroe	1937 & 49	——	Newark, NJ
HULSANDER ENG. CO.	Harris St.	1944 & 49	——	Trenton, NJ
HYATT MFG. CO.	1 N. Holiday	1937 & 39	——	Baltimore, MD
H.W. PETERS INC.	5178 Wash.	1939	——	Boston, MA
H.W. SCHWAB, INC.	43 Union Square	1939	——	New York, NY
I.B. & CO.	——	Unkn.	City HM	Philadelphia, PA
INDIANAPOLIS BADGE	3823 Mass. Ave.	1949-1991	IN, Address HM	——
INDIANAPOLIS BADGE	25 W. McLean Pl.	1992-Present	IN, Address HM	——
INGRAM-RICHARDSON MFG. CO. Oregon	1937	——	Beaver Falls, OR	
INGRAM-RICHARDSON MF	1938 McMillan	1939	——	Beaver Falls, OR
IRVINE & WIRTH & J.	2439 Market	1906-1907	Temp/Earthquake	San Francisco, CA
IRVINE & WIRTH & J.	2129 Market	1907-1910	See J.C. IRVINE	San Francisco, CA
IRVINE & JACHENS	2129 Market	1909-1910	Wirth Died/New Co.	San Francisco, CA
IRVINE & JACHENS	1027 Market	1910-1925	Larger Bldg.	San Francisco, CA
IRVINE & JACHENS	1068 Mission	1926-1950	Address/City HM	San Francisco, CA
IRVINE & JACHENS	6700 Mission	1951-Present	Daly City, CA	San Francisco, CA
IRVIN HAHN CO.	Unkn. Starting Ad	1890-Unkn	——	Baltimore, MD
IRVIN HAHN CO.	207 Sharp	1937 & 39	——	Baltimore, MD
IRVIN H. HAHN CO	1830 Worcester	Unkn.-Present	——	Baltimore, MD
IRONS & RUSSELL CO.	——	1937 & 39	Listed Co.	Providence, RI
IRONS & RUSSELL CO.	95 Chestnut	1944	——	Providence, RI
IRWIN	——	Unkn.	Name HM	Portland, OR
IRWIN-HODSON CO.	——	1917-1970	City/State HM	Portland, OR
IRWIN-HODSON CO.	378 Washington St.	1895-1900	Badges perhaps	Portland, OR
IRWIN-HODSON CO.	401 Broadway	1930-1934	——	Portland, OR
IRWIN-HODSON CO.	439 N.W. 15th St.	1935-1970	——	Portland, OR
IRWIN-HODSON CO.	15th & Glisan	1937 & 49	——	Portland, OR
IVER JOHNSON CO.	——	1880-1900	City HM	Boston, MA

(Image 257)

Considerations in Attempting to Date a Particular Badge

Unless a USN police-style badge is coming from a sailor's documented estate or carries other valid documentation, then it is difficult to date. Researching a listed unit name (e.g. USS Coral Sea) will allow a beginning and ending date range for usage of that badge; additionally, if the badge has a ship's designation as part of the badge (e.g. CVB-43), a review of that ship's history might show that the unit designation was changed to CVA-43 on such-and-such date. That would further limit the valid usage date for that badge. Note that the Navy is well known for allowing a "wear out period" whenever implementing uniform changes, and the MAA Force of CVB-43 would most likely continue wearing the original badge past the unit designation change date. But any new badge orders would certainly say, in this example, CVA-43 on the badge and not CVB-43 as on the original badge.

Most naval training centers (NTCs) were originally called naval training stations (NTSs), and those designations on badges will help limit a valid usage date range.

There are a number of things to check in attempting to date a particular badge; however, even with those checks it is only possible to positively date a badge if full provenance is included with a badge. For example, the original owner sells you the badge and gives the history of that badge.

Keep in mind that other variables can throw off dating a badge. For example, some vendors had badge blanks (badges complete except for engraving title/unit/etc.) in their inventory for up to 20 years or more. Secondly, broken badges might have had the badge catch replaced by a current style catch that can change the look of an older badge to look like something 20 or 30 years newer or older.

Even without full provenance for a badge it is possible, but not definitive, to check the points listed below and possibly narrow the valid date range of a badge:

a) Type of catch (evolution of catch designs with <u>previously noted caution</u>)
b) Hallmark (company date history, see previous Monty McCord/Stephen Knight info)
c) Unit listed on badge (unit date history, unit name change, ports of call)
d) Former service member associated with a particular badge (for example, Petty Officer "Ace" Vokoun was stationed aboard the USS Courtney from 1961-63, and you are considering purchasing a badge marked "USS Courtney" from Vokoun's family).

Fakes

Certain examples of vintage police badges, including military versions, have become so valuable that individuals and even foreign companies have begun producing fake copies. If the provenance (documented or undisputed proof of ownership) of a piece you are interested in is not conclusive, get a second opinion from a seasoned collector before purchasing.

Nicely done cast reproduction of a rare MAA badge
(Image: 258)

Cast badge, USS Sierra crest applied over a production LE crest
(Image 259)

U.S. Navy crest over a generic LE crest (Image 260)

Error Badges

Some commands utilized badges even though errors had been made during production. Here are two examples.

US(S) Frederick (Image 261)

USS Bronstein (DE-1037) (Image 262)

Repurposed Badges

MAA badge from Pearl Harbor base unit, repurposed for an aviation Unit (VR-8); moved to Hickam AFB (not Pearl Harbor), Hawaii, in 1949; moved to Moffett Field, Calif., in 1957. (Image 263)

Other U.S. Service Branch Examples

The U.S. Army established a Provost Marshal as early as 1776 and a Provost Corps in 1778. The Provost Corps was the forerunner of the Military Police Corps, but after each major American war from 1776 through 1918, the Provost Corps was disestablished.

A service school for military police was not established until WWI. This school was established in Europe, and the first chief instructor was a British officer. After WWI, the military police only officially existed in U.S. Army reserve units. On September 26, 1941, the Secretary of War established a permanent Army Military Police Corps.[xiv]

Image 264	Image 265
Shape: Stylized eagle-top shield	Shape: Eagle-top pinched shield
Notes: Provost Investigator; all Army MPs fall under the purview of the Provost Marshal.	Notes: Early Military Police, 1920s Panama Canal.
Images:	Images: FlyingTigerAntiques.com

Image 266	Image 267
Shape: Eagle on sunburst shield	Shape: Custom
Notes: U.S. Army Military Police school	Notes: U.S. Army Military Police; 2" reserve officer's overseas cap badge; described in orders establishing 1920s Army Reserve MP Corps. Extremely rare.
Images:	Images: Bill Hedges collection

Image 268	Image 269
Shape: 6-Point, ball-tipped star	Shape: Eagle-top shield and oval
Notes: USMC Military Police	Notes: USMC “Marine Police” and Army Mil. Intelligence, 1930s
Author’s collection	Images: FlyingTigerAntiques.com

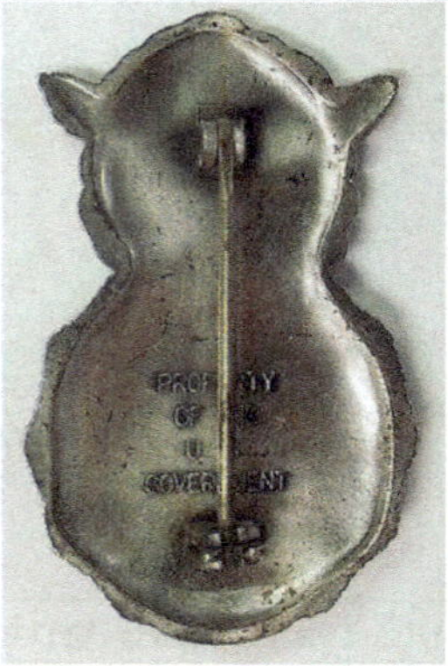

Image 270	Image 271
Shape: Eagle-top shield	Shape: Eagle-top pinched shield
Notes: U.S. Air Force, early “Air Police,” pre-1966.	Notes: Likely 1940s to early 1950s; one of the three USAF bases in Albuquerque, NM.
Images:	Images:

Image 272	Image 273
Shape: Pinched shield	Shape: Sunburst and teardrop
Notes: U.S. Coast Guard port security	Notes: USCG, MAA and OOD
Images:	Images: Author’s collection (MAA)

Civilian Law Enforcement Badge Examples

Image 274	Image 275
Shape: Shield w/cut-out star & shield	Shape: Pinched shield & "radiator" shield
Images: Dave Snyder collection	Images:

Image 276	Image 277
Note: Very rare and valuable jeweler made "presentation" badges.	Note: Jeweler-made "presentation" badges, including highly desirable "suspension" badge
Images: James C. Casey	Images: James C. Casey

Image 278	Image 279
Images:	Images:

Badge Display Possibilities

Semi-fancy displays (left) and simple, inexpensive Riker Mounts display cases (right) (Image 280)

James (Jim) Casey with award-winning traveling displays (Image: 281)

Possible Future Directions of Collecting USN-Related Memorabilia

Here are some items that may have future collecting interest:

a) Duty belts
b) Leggings
c) Batons and baton holders
d) Brassards
e) Whistles
f) Handcuffs (Restraints)
g) WWII and earlier naval base civilian worker ID badges
h) Shore patrol vehicles (Jeeps, motorcycles)
i) Full uniforms
j) Patches
k) USN firefighting memorabilia

l) WWII-era "sweetheart" pins (Image 282)

Tips on Collecting U.S. Navy Law Enforcement-Style Badges

Collector clubs can assist in collecting police badges in general, and police badge collectors occasionally offer military police and USN Master-at-Arms badges. The author has found that serious police badge collectors sometimes have sub-collections of military police and Navy police badges. In addition, these collectors are often less attached to their sub-collections and more willing to sell or trade those badges.

With careful watching to avoid obvious fakes, you can find good U.S. Navy law enforcement-style badges from some sellers on eBay. An estate collection will occasionally be presented for sale on eBay. Two of the cover badges from this book were purchased from an estate sale on eBay. (Three other eBay badge purchases turned out to be faked.)

A good starting place for collecting these badges could be an online forum such as the United States Militaria Forum (USMF). Try placing a post stating that you collect military-related police-style badges to see if other collectors answer your post with helpful information.

U.S. Navy law enforcement-style badges are "cross-collectibles." They are sought by past members of military units who are looking for a memento from their service days as well as overall collectors of militaria. In addition to militaria collectors, Navy police-style badges are a global collectible. They are sought by police memorabilia collectors who are willing to pay top dollar for the more desirable badges.

If you choose to collect items covered in this book or items from possible future collecting directions, I wish you luck in your efforts.

GLOSSARY

Badge layers: Any layer added after the initial die-stamped blank is produced (e.g. a crest)
Badge shape: Shield, oval, star, and round are the four most basic badge shapes
Ball catch: Burgess catch with a fully rounded "ball"-shaped head
Ball-tipped: Round decorations added to the points of star badges, either stamped or applied
Bar pin: Flat bar used to attach the pin and catch as a single piece; the first style patent was granted in January 1937, and the second style was granted in August 1945
Brass: Base material used in the production of many badge blanks
Burgess catch: A popular style of clasp catch patented in July 1909; small or large Burgess found
C-catch: Early catch style, late 1800s-early 1900s; catches the pin end and looks like an open "C"
Catalog/Stock badge: Inexpensive stocked badges for sale by most early makers; additional lettering could be added for an extra cost
Catch/Clasp: Some collectors use the terms interchangeably
Chrome: A final badge plating that is less expensive than nickel but looks shinier than nickel
Dapping: Manually turning a flat piece of metal into a curve using dies and punches
Die-cut: Either a standard or a custom machine die used to produce a stamped badge blank
Eagle top: Any badge that incorporates an eagle at the top of the badge design
Fork catch: Clasp catch, a.k.a. side swing catch (early to mid-1900s usage)
Hand engraved: Letters and numbers added to a badge blank by hand stamping or engraving
Jeweler badge: A more expensive, custom badge produced by a jeweler; mostly civilian LE usage and often from precious metals
Law Enforcement (LE): (or **LEO**) Officers enforcing state, federal, tribal, military and local laws
Machine engraved: Letters and numbers are added to a badge blank by a machine
Nickel: Most common badge final plating
Oval: One of the four most basic badge shapes
Panel / Scroll: A small strip of metal added to the badge blank used to hold additional lettering
Precious metal: Gold, silver, or platinum content used in badge or jewelry-making
Presentation badge: Civilian LE badge, mostly jeweler-made, presented to a retiring law officer
Round: One of the four most basic badge shapes
Saddle: An extra thin strip of metal used to help secure a badge's pin mechanism
Seal / Crest: A Center emblem added to a badge, sometimes denoting the unit, most often generic
Screw post: A small upside-down bolt attached to the badge with a round "nut" for attaching a badge to a hat
Shield: One of the four most basic badge shapes
Spring pin: Badge clasp coiled at one end, like a safety pin coil, produces a spring at the free end
Stamped: Physical method of making a die-cut badge blank
Star: One of the four most basic badge shapes
Sterling: A mark denoting at least 92.5% silver content in badges and jewelry
Suspension badge: Early civilian lawmen sometimes wore a two-piece badge with the two parts connected by small chain links
Teardrop shield: Eagle-topped shield shape that some vendors say resembles a teardrop
Tin star: Early frontier lawmen sometimes produced a temporary badge from tin cans
Tongue catch: Like a "C" catch but much wider and flatter; it looks like one's tongue stuck out
Vaulted: Term for a reverse-stamped curved badge, which can be either a slight vault or a heavy vault
Wire pin and catch: Single piece of wire bent back and with a built-in "C" catch

RESOURCES

BIBLIOGRAPHY

Bureau of Naval Personnel, 1943, U.S. Navy Shore Patrol Manual, (Navy Dept., DC)
Bureau of Naval Personnel, 1953, rev.1964, The U.S. Navy Shore Patrol, NAVPERS 10059-B, (Naval Training Publications Center, Washington, DC)
Bureau of Naval Personnel, 1953, U.S. Navy Shore Patrol Manual, (Navy Dept., DC)
Bureau of Naval Personnel, 1966, U.S. Navy Shore Patrol Manual, (Navy Dept., DC)
Casey, James C., 2008, Badges of America's Finest, (M.T. Publishing Co., Evansville, IN)
Hedges, William B. and Baird, L. Lawrence, 1993, Badges A Guide For The Serious Collector, (L.L. Baird, Moreno Valley, CA)
Hedges, William B., 1992, A History of United States Military Law Enforcement and Insignia, (Bill Hedges, Buena Park, CA)
Lucas, Kenneth W., 2010, Military Police Badges, (K. Lucas, Chesapeake City, MD)
McCord, Monty, 1999, Law Enforcement Memorabilia, (Krause Publications, Iola, WI)
Smith, Daniel Sr., 2012, U.S. Navy Sailors' Memorabilia, (D. Smith, Chattanooga, TN)
Stacey, John A., 2008, United States Navy Rating Badges and Marks (1833 to 2008), (J. Stacey, Nanjemoy, MD)
U.S. Navy, 1939, *The Bluejacket's Manual* (United States Naval Institute, Annapolis, MD)
U.S. Navy, multi years, *Uniform Regulations (NAVPERS 15665I)* (Washington, DC)
Virgines, George, 1987, Badges of Law and Order, (Cochran Pub. Co., Rapid City, SD)
Virgines, George, 1989, Police Collectibles Pictorial Guide, (Baird Pub. Co., Moreno, CA)

(This bibliography is not an exhaustive list of published works on LE badge collecting)

WEB RESOURCES

All Hands Magazine; back issues, multi years, Navy.mil/allhands.asp?
Bluejacket.com; bluejacket.com; excellent USN reference site
CollectorsWeekly.com/search.php?q=police+badges
FlyingTigerAntiques.com; top quality military badges of all types
James C. Casey; PoliceGuide.com
Naval Historical Center; History.Navy.mil
NavyCollector.com; Image galleries of USN collectibles
U.S. Militaria Forum (USMF): USMilitariaForum.com; covers all aspects of U.S. military collecting

ENDNOTES

[i] George E. Virgines, Badges of Law and Order, (Cochran Publishing Co., Rapid City, South Dakota, 1987)
[ii] Wikipedia, retrieved on January 7, 2016.3
[iii] Timothy Roufa, CriminologyCareers.About.com, retrieved on January 7, 2016.
[iv] James C. Casey, www.PoliceGuide.com/timeline, retrieved on February 10, 2016
[v] John A. Stacey, United States Navy Rating Badges and Marks 1833 to 2008, (J. Stacey, Nanjemoy, MD, 2008)
[vi] Navy MAA website, https://www.navycs.com/navy-jobs/master-at-arms.html
[vii] James C. Casey, www.PoliceGuide.com, retrieved on March 9, 2016

[viii] The United States Naval Institute, The Bluejackets' Manual, (Annapolis, MD, 1968)
[ix] Albert L. Raithel, Jr., Capt., USN(Ret.), Naval Aviation in WWII
[x] Bureau of Naval Personnel, U.S. Navy Shore Patrol Manual, (Navy Department, Wash., DC, 1943)
[xi] Bureau of Naval Personnel, U.S. Navy Shore Patrol Manual, (Navy Department, Wash., DC, 1953)
[xii] George E. Virgines, ibid.
[xiii] George F. Virgines, ibid.
[xiv] William B. Hedges, A History of United States Military Law Enforcement and Insignia, (Pub. By the Author, Buena Park, CA, 1992)

About the Author

Daniel Dee Smith, Sr.

Author and Military Historian Daniel Smith, Sr. is retired from twenty-six years in the U.S. Navy. After retiring from the Navy, he joined the all-volunteer Tennessee State Guard as a Captain (Infantry). He retired from the State Guard after seven years as a Lieutenant Colonel.

He has an undergraduate degree from Covenant College and performed post-graduate work in World History at California State University.

After retirement from the Navy, he spent 23 years as an IT manager at the Tennessee Valley Authority and another ten years as a consulting Senior Project Manager at Olav & Co.

He was an associate editor of the automobile magazine *Vette Vues* in the mid-1970s.

From 1997 to 1999, Smith was executive director of the National Medal of Honor Museum of Military History in Chattanooga, Tennessee.

He has published four military history reference books, many classic car magazine articles, and short story fiction pieces. His military history books include three on U.S. Navy memorabilia and one book titled "*The Medal of Honor and the Battles for Chattanooga*."

He specializes in the history of the military Medal of Honor and World War II military history.

Aside from reading, his favorite pastime is showing and judging classic automobiles.

He and his wife, Betty Thomas Smith, have two grown children and reside in Chattanooga, Tennessee.

Visit at DanielSmithBooks.net

Made in the USA
Monee, IL
02 November 2026

67378329R00050